W9-BMW-034

# www.wadsworth.com

*wadsworth.com* is the World Wide Web site for Wadsworth Publishing Company and is your direct source to dozens of online resources.

At *wadsworth.com* you can find out about supplements, demonstration software, and student resources. You can also send e-mail to many of our authors and preview new publications and exciting new technologies.

**wadsworth.com**
Changing the way the world learns®

# The Counselor Intern's Handbook

## SECOND EDITION

**CHRISTOPHER FAIVER**
John Carroll University

**SHERI EISENGART**
John Carroll University

**RONALD COLONNA**
Neighboring Mental Health Services

**Brooks/Cole**
Thomson Learning™

Australia • Canada • Denmark • Japan • Mexico • New Zealand • Philippines
Puerto Rico • Singapore • South Africa • Spain • United Kingdom • United States

Executive Editor: *Craig Barth*
Counseling Editor: *Eileen Murphy*
Assistant Editor: *Julie Martinez*
Marketing Manager: *Jennie Burger*
Project Editor: *Howard Severson*
Print Buyer: *Stacey Weinberger*

Permissions Editor: *Susan Walters*
Copy Editor: *Meg McDonald*
Cover Designer: *Carole Lawson*
Compositor: *Scratchgravel Publishing Services*
Printer: *Malloy Lithographing, Inc.*

COPYRIGHT © 2000 by Wadsworth,
a division of Thomson Learning

Brooks/Cole Counseling is an imprint of
Wadsworth

All rights reserved. No part of this work covered
by the copyright hereon may be reproduced or
used in any form or by any means—graphic,
electronic, or mechanical, including photo-
copying, recording, taping, or information
storage and retrieval systems—without the
written permission of the publisher.

Printed in the United States of America
2   3   4   5   6   7   03   02   01   00

For permission to use material from this text,
contact us:
   **Web:** www.thomsonrights.com
   **Fax:** 1-800-730-2215
   **Phone:** 1-800-730-2214

**Library of Congress
Cataloging-in-Publication Data**
Faiver, Christopher, 1947–
   The counselor intern's handbook /
Christopher Faiver, Sheri Eisengart, Ronald
Colonna. — 2nd ed.
      p.   cm.
   Includes bibliographical references and
index.
   ISBN 0-534-24882-9 (pbk.)
   1. Counseling—Study and teaching
(Internship)—Handbooks, manuals, etc.
I. Eisengart, Sheri, 1949–   . II. Colonna,
Ronald, 1950–   . III. Title.
BF637.C6F32   2000
158'.3'07155—dc21                99-29251

**Wadsworth/Thomson Learning**
**10 Davis Drive**
**Belmont, CA 94002-3098**
**USA**
**www.wadsworth.com**

**International Headquarters**
Thomson Learning
290 Harbor Drive, 2nd Floor
Stamford, CT 06902-7477
USA

**UK/Europe/Middle East**
Thomson Learning
Berkshire House
168-173 High Holborn
London WC1V 7AA
United Kingdom

**Asia**
Thomson Learning
60 Albert Street #15-01
Albert Complex
Singapore 189969

**Canada**
Nelson/Thomson Learning
1120 Birchmount Road
Scarborough, Ontario M1K 5G4
Canada

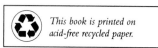
*This book is printed on
acid-free recycled paper.*

# Contents

## 6    PSYCHOLOGICAL TESTING OVERVIEW    56

## 7    INTEGRATING PSYCHOTHERAPY AND PHARMACOTHERAPY: GUIDELINES FOR COUNSELOR INTERNS    71

# Preface

This basic guide is designed to assist counseling students and others in the helping professions through the entire experience of internship, often the last requirement of a degree program. We suggest its purchase before the internship placement because Chapter 1 gives suggestions for selecting a site. Further, many will find it helpful as a study guide for licensure and certification examinations, and seasoned professionals may use it as a quick reference. We include overviews of basic treatment modalities, psychological testing, and psychopharmacology, as well as chapters on the clinical interview and ethical considerations—subjects that are particularly relevant to interns. Our fields are counseling and psychology, yet students in related areas of the behavioral sciences may find the book helpful.

Among us we have supervised field placement students in a variety of settings for more than 50 years, so we are able to maintain a practical perspective in this book. Our varied backgrounds include teaching, administration, and clinical work. We encourage students to review what they have learned while they progress to new material. We explain the components of internship in detail, and we urge students to use them with enthusiasm as well as caution. It is important to leave no stone unturned during the internship process, but we provide only the stones, respecting the student's responsibility for the turning.

## ACKNOWLEDGMENTS

Some of the information we offer emerged from institutions involved in providing field experiences to interns; in particular, we thank the John Carroll

University Counseling and Human Services Program Committee. A number of students and colleagues contributed valuable assistance while we were writing this book. They include Paula Britton, Ph.D.; Geriann Cahill; Matthew Guzzo; David Helsel, Ph.D.; Valerie Moyer; Pamela Nims; Eugene O'Brien, Ph.D.; and John Ropar, Ph.D. We appreciate the thoughtful comments of Richard Schwartz, M.D., Case Western Reserve School of Medicine, regarding our psychopharmacology chapter. We also appreciate the efforts of the following reviewers: Beverly B. Palmer, California State University—Dominguez Hills; Myrna Thompson, Southside Virginia Community College; Daisy B. Ellington, Wayne State University; C. Emmanuel Ahia, Rider University; and Cindy Juntunen, University of North Dakota.

The staff at Brooks/Cole•Wadsworth was enormously helpful. We appreciate the encouragement and support of Eileen Murphy and the work of production editor Howard Severson and copy editor Meg McDonald.

We look forward to your comments. Please write to us at Brooks/Cole• Wadsworth Publishing Company, 511 Forest Lodge Road, Pacific Grove, California 93950.

*Christopher Faiver*
*Sheri Eisengart*
*Ronald Colonna*

**1**

# Getting Started

Your internship is generally the culmination of the academic sequence leading to your degree in counseling. It is an exciting, challenging time, and many students anticipate it with mixed emotions. Internship students often find themselves facing unfamiliar situations, engaging in intense encounters, and processing powerful feelings, which all lead to increased introspection, personal reassessment, change, and growth. The internship may be your first opportunity to interact independently with real clients in a professional capacity as a counselor, as well as your first experience in scrutinizing those interactions during a formal supervisory hour. This book is a step-by-step guide that will remove some of the anxiety from your internship and allow you to relax and enjoy the many rewarding, wonderful moments you are about to experience as a counselor intern.

Ideally, your internship should give you a supportive, structured learning environment for acquiring clinical experience and practical on-the-job training. You will be called on to synthesize material from previous coursework, to utilize theories and techniques, and to begin to develop a personal and professional style of relating effectively to clients, clients' families, agency staff members, and other mental health professionals.

During your counseling internship, you will work under the direct supervision of a licensed counselor, social worker, psychologist, or psychiatrist. You will meet for regular supervisory sessions to review your internship experience, as specified by your state licensure/certification board. Ideally, your field placement supervisor should be an experienced clinician who will provide a flexible learning experience tailored to meet your individual needs by encouraging you and continually assessing your strengths and weaknesses as well as asking for your feedback. The actual number of hours you spend in direct contact with clients, as well as the total number of supervisory hours, should always fulfill state licensure/certification requirements. For example, the Council on Accreditation of Counseling and Related Educational Programs (CACREP) specifies that 40 percent of the required 600 internship hours should be spent in direct client contact and that the intern should have one hour of supervision for every 20 hours of work. So a graduate student fulfilling a 600-hour internship would need 240 hours of client contact and 30 hours of supervision to have the internship meet these national standards. Many states have adopted these standards for licensure in counseling.

In addition to on-site experience and supervision, you will most likely meet on campus with other interns to discuss your placement experiences. Most students appreciate this group time with their peers to share their feelings, their frustrations, and their accomplishments, and they find the support this group offers especially helpful during their internships. You may also meet with a supervisor from your counseling program (usually one of the department or program professors) for further processing of your individual experiences. You will typically be required to keep a detailed log or journal of daily on-site activities and to prepare two or more formal case studies for class presentation. Additional research projects or reading relevant to counseling may be assigned, as well. The agency and university supervisors will evaluate your performance and potential as a counselor, and they should discuss their evaluations with you both midway through the internship and again at completion of the experience.

## SELECTING YOUR INTERNSHIP SITE

Selecting your placement site is the first, and perhaps the most important, step in your internship experience. The selection process integrates four factors:

- Your own interests and needs
- The field placement guidelines of your university counseling or human services program
- The state requirements for on-the-job experience for professional counseling licensure/certification
- The didactic and experiential opportunities afforded at the placement site

These four factors are interdependent, and selecting a placement site is a dynamic process of exploring and matching these various criteria to find a good fit. Some counseling and human services programs have formal placement relationships with preapproved agencies, which may limit selection to a certain extent; but this may also help provide a satisfactory field experience. Other programs delegate most of the responsibility for securing field placements to the individual student. In either case, you should first try to determine your own interests, needs, and expectations when you begin the process of choosing a placement site. (A few programs assign interns to a specified site with little or no input on the part of the student.)

## Your Interests and Needs

You may already have a fairly good idea about what types of people you enjoy working with (for example, children, adolescents, adults, or the elderly) and what kinds of problems you would like to deal with (for example, substance abuse, child and family concerns, career counseling, or mental health issues such as depression and anxiety). Knowledge about your own preferences gives you some direction when you begin looking for an internship site because you can limit your choices to those places where you are certain that you have a keen interest in agency clients and services. You may also use your internship as an opportunity to try something new, to deepen your self-awareness, and to enlarge your scope of experience. For example, if you have been working with abused children but have not had any helping interactions with chemically dependent adults, your internship may provide a chance to obtain practice in a different area.

Agencies can have client types and services that are either homogeneous (such as drug and alcohol treatment centers or child welfare agencies) or heterogeneous (such as community mental health centers or psychiatric hospital units). If you are unsure about your interests, or if you do not prefer working with one type of client or one particular type of problem, then you may want to consider heterogeneous internship sites, which will offer you a wider range of experience. We suggest that you consider the opportunity of working with persons from a wide variety of cultures and ethnic backgrounds.

In beginning the selection process, you should also consider whether an inpatient or an outpatient setting would satisfy more of your needs and interests. The hospital inpatient setting will typically be fast-paced and high-pressured, with a rapid client turnover. If you select a psychiatric inpatient unit for your internship, you will most likely come into contact with a great many clients, most of whom are in acute distress and are manifesting serious psychopathology. Working with the psychiatric inpatient is similar to crisis intervention in that your counseling relationship and interventions help the client stabilize and return to a prior level of functioning. The hospital unit lets you interact with other health and mental health professionals as a member of an interdisciplinary treatment team, as well as learn to relate to clients who have complicated psychiatric disorders, and these experiences are valuable and interesting. However, the intensity of the emotional upset and the extent of the problematic behaviors encountered in a hospital inpatient unit may not suit all counseling students.

Long-term residential treatment facilities provide clients who generally have chronic, rather than acute, problems or difficult management issues and who require a more structured or supportive environment. One advantage of this type of internship is the extended counseling relationships that you may develop with your clients, who are likely to remain at the facility throughout your placement. In addition, you may have special opportunities to observe, learn about, and interact with your clients as they go about their daily activities. An internship at a long-term treatment center for children or adolescents can be particularly rewarding in these respects.

If you choose an outpatient setting, your internship experience will vary according to the particular agency. Some agencies may let you work as a member of an interdisciplinary team and participate in treatment conferences, whereas at other agencies you may interact only with your supervisor to discuss cases. Most community mental health centers must by law provide services in such areas as intake assessment, emergency care, consultation and education, research and evaluation, individual and group counseling, and after-care planning. The outpatient setting will usually, but not always, serve clients who are less distressed and less acute than those in the inpatient unit. As an intern in an outpatient center, you may have more opportunities to use a variety of counseling techniques because the clients are generally higher functioning and are able to cope more effectively with daily living tasks. However, if you work in the intake or emergency department of a community mental health center, you may deal with highly upset clients or clients in crisis, who may require immediate hospitalization. In addition, many of your clients may have chronic problems or be "repeaters" who require continued support and management.

Another aspect of your internship needs may involve financial considerations. At some agencies interns are paid for their services; however, these sites will be limited, and therefore your choices will be somewhat restricted if you require financial compensation during your internship. Unfortunately, many excellent internships are unpaid positions. Some agencies will hire a graduate counseling student to perform a paid job, such as case manager or mental health technician, and then allow the student to set aside a designated number of unpaid hours specified as internship hours, when the student assumes tasks and responsibilities relevant to the internship. Some students find this arrangement satisfactory, whereas others report that they are frustrated with their limited counseling and supervised time.

You should also consider whether a particular agency's schedules mesh with your own needs. For example, some agencies offer group therapy sessions on several evenings during the week, which may conflict with your personal responsibilities.

## Your Counseling Program Guidelines

As you begin exploring possible internships, be certain that your potential site fulfills all requirements of your counseling program and state licensing board. Most programs print clear guidelines listing the expectations and regulations for student internships. For example, your counseling program guidelines may specify that interns acquire experience in treatment planning, in individual and group counseling, in case management, and in discharge planning, as well as gain an understanding of agency administrative procedures. You should be sure that you will be able to satisfy all the necessary program requirements at each internship site. In addition, you must determine whether you are covered by malpractice insurance provided by your program. Your counseling program guidelines should specify whether students are covered. If you need to purchase liability insurance, professional organizations such as the American Counseling Association (ACA) provide coverage at reduced rates for students.

## State Licensure/Certification Requirements

Many state counselor licensure/certification boards allow credit for supervised hours accrued during your internship, even before you complete your graduate program in counseling. However, specific rules and regulations address, for example, whether your experience may be paid or unpaid, the hours and nature of supervision, the relationship between supervisor and supervisee, and the intern's scope of practice. In addition, state boards have formal procedures and policies for registering supervised counseling experience. As soon as you begin selecting an internship site, you should write or call your state counseling regulatory board to obtain all documents, applications, and instructions pertaining to counselor licensure and certification. Then read through everything carefully and consider whether you will be able to fulfill the state requirements for supervised counseling experience at each internship site. In addition, once you have selected a site, be sure to follow through on all formal procedures to register your hours with your state board so that you can get credit toward state licensure or certification for your internship.

## Experiential and Didactic Opportunities of Each Site

Your internship is a critical part of preparing for your career as a professional counselor, and you should carefully examine and analyze what kinds of educational opportunities you will be likely to experience at each potential site. Will you have direct client contact, such as doing intake assessments and conducting individual or group counseling, or will much of your time be spent running errands, such as filing papers and operating the copying machine? Will you be treated as a colleague and a valued member of the treatment team, or will the other professionals discount your input because you are a student? Will you have the support you need, and will other staff members be willing to answer questions and offer help if you ask? Will you be invited to attend staff meetings and in-service educational sessions?

Your prospective supervisor will be responsible, in large part, for delineating your internship activities. Therefore, you should thoughtfully consider the personal and professional qualities you hope to find in this individual. Your supervisor also must have the time, the interest, and the commitment to teaching interns, plus an understanding of the special nature of the supervisory relationship.

## SOURCES OF INFORMATION

As you begin the process of selecting your internship, you will need to assemble a list of potential sites. You may already have some thoughts about where you would like to do your internship. However, if you have no idea where to begin, we suggest a preliminary discussion with your counseling program coordinator of field placements. The coordinator should be able to provide helpful suggestions and insights concerning prospective placement sites, as well as names of alumni of your program who are currently working at various sites who would be good people to contact for further information. It may also be a good idea to sit down with your faculty adviser, who ideally knows something about your personality and needs, and who may be able to recommend several possible placements. Many universities have a list of approved internship sites.

Networking with other students who are currently involved in a field placement or who have recently completed one is an invaluable way to gather information about potential sites. In this way you may hear of several interesting sites that you did not know were available. You will probably also learn a great deal about a particular site's activities, schedule, staff, and atmosphere by talking to students who have worked there. Some of these same questions may be answered by site supervisors at the informational interview, which is discussed in the next section; however, it is often helpful to have another student intern's perspective as well.

One additional source for potential field placement sites is the classified ad section of your local newspaper. The help wanted advertisements, under such headings as "social services," "medical settings," "counselors," and "mental health," often provide possible leads. You can telephone each agency and ask whether an internship or field placement program is available and whom to contact for further information. In addition, many funding agencies, such as the United Way or local mental health board, publish listings of their affiliated agencies. Often these listings, which usually include agency service descriptions as well as contact persons, provide good leads for internships.

## INFORMATIONAL INTERVIEWING

You may acquire awareness of your own needs, interests, and expectations through course readings and experiences, as well as through actual visits to potential placement sites. Discussions with supervisors and clinicians at the site will enable you to clarify the four factors to consider in choosing your internship: (1) your own needs and interests, (2) your counseling program guidelines, (3) state regulations concerning counseling licensure/certification, and (4) the educational opportunities at each site. A visit to a potential site helps you determine whether that particular internship site is a good fit and meets most of your expectations. This process, known as informational interviewing, can give you invaluable firsthand exposure to each site's unique environment.

To prepare for an informational interview, we suggest that you compose or update a resume outlining your educational and professional experience, including volunteer activities relevant to counseling or human services. (A sample resume is provided in Appendix A.) Take along a copy of your resume, plus any appropriate samples of your written work (such as case studies or research projects from courses), and also a copy of your counseling program's field experience regulations, including agency guidelines, course requirements for students, and evaluation procedures.

You should be prepared to answer, as well as to ask, questions when you go on your informational interviews; the professional at the placement site may look on this interview as an ideal opportunity to learn something about your personality and ability. Future supervisors have asked potential student interns a wide variety of questions, ranging from "What is your theoretical orientation?" to "What is your favorite restaurant?" to "Have you ever been in therapy yourself?" Your attitude toward being questioned and your style of relating to the interviewer may be more important than the actual answers to those questions.

Following the interview, we suggest that you send a brief thank-you note indicating your interest, if any, in the placement and noting that you will follow up with a phone call in one or two weeks to explore the possibility of arranging an internship at that site. (Appendix B provides a sample thank-you note.)

## JOB DESCRIPTIONS

Once you have listed potential placement sites and have visited the most promising choices, you can compare aspects of each placement to find one that best meets most of your needs and interests, fulfills your academic program specifications, satisfies as many of the state licensing/certification requirements as possible, and provides the optimal educational opportunities. To facilitate selection, you may find it helpful to prioritize criteria. For example, you may decide that it is more important to you to work with one specific client population than it is for you to work as an interdisciplinary team member. In addition, keep in mind that the goal is to find a good fit, rather than the perfect choice, so try to be flexible in assessing each placement site.

Job descriptions provide key information for you to consider, in conjunction with other data, because they represent the supervisor's or administrator's standards and expectations for each placement experience. You may be able to pick up printed job descriptions in the program office at your college or university, in the classified ad section of the newspaper, or in various agency offices. You may also find it helpful to write notes on the verbal job descriptions given at informational interviews or during discussions with other students, advisers, or professors. These notes can be reviewed and compared later on. (A sample internship announcement is provided in Appendix C.)

As you review these job descriptions to select your internship site, you may find it helpful to consider the following questions:

- What kinds of people do I enjoy working with? Will I be likely to work with this population at this placement site?
- What types of problems might I want to deal with? Will I encounter these types of problems here?
- Is the experience paid or unpaid?

- Does the placement allow for credit toward state licensure or certification requirements? Some state licensing/certification boards, for example, specify that the student must receive at least one hour of supervision for every 20 hours of client contact.

- How much will I actually be working with clients, rather than observing or running errands?

- Will I receive adequate individual supervision? Is the supervisor credentialed at the independent practice level of licensure or certification? Having met the supervisor, do I feel that we will be able to maintain a good working relationship?

- Am I covered by liability insurance? Most institutions of higher education carry policies that insure students during their internships. In cases where the university or college does not provide insurance, students should seek coverage through other sources, such as professional organizations (for example, the ACA).

- What is the general atmosphere at this placement site? Is it formal or informal? Will I be comfortable working here? Will I be welcomed as a colleague?

- What other mental health professionals are on staff? Will I be directly involved with psychiatrists, psychologists, social workers, counselors, nurses, teachers, or child care workers? Will I be part of a treatment team?

## GETTING READY FOR THE PLACEMENT

Finally! You think you have found an internship site that is a good fit for you, and your future supervisor has told you that you have the position. You wonder what to do next. First, you will need a written agreement for the placement, signed by your future supervisor, specifying the dates of your internship. Next, you should follow the prescribed state licensing/certification procedures to obtain credit for your supervised hours. This process usually involves both you and your supervisor completing paperwork describing the proposed scope of practice, the setting, the number of hours of work and of direct supervision, and the supervisor's areas of expertise.

Your meeting with your supervisor, when these official forms are completed and signed, is a good time for you to ask, "What can I do, before I actually begin work, to best prepare myself for this placement?" Your supervisor may be able to offer some advice, for example, by suggesting a few books for you to read or by noting that you should thoroughly review the *Diagnostic and Statistical Manual of Mental Disorders* before beginning the placement experience. At this meeting with your supervisor, you may also want to discuss the actual scheduling of your work hours so that an agreement acceptable to both of you can be worked out. Some students have found it useful to discuss dress codes, if applicable, and whether they will need to bring any special equipment or supplies. For example, one inpatient unit we know requires all counselors to use pink highlighters on their patient chart notes to make these notes easy to see. At this internship site, interns come prepared with several pink highlighters each day.

You may also find it helpful, during the time before you begin your internship, to review academic coursework, concentrating on those areas you feel might require extra attention or study. The National Board of Certified Counselors (NBCC) provides coursework descriptions in the 10 areas of study relevant to counseling, and this may be used as a guideline for self-assessment (see Appendix D for the NBCC course descriptions). We also suggest a thorough reading of the American Counseling Association Code of Ethics and Standards of Practice (see Appendix E) so that you will be ready to act in a professional manner in all situations.

## OTHER PLACEMENT VENUES

Although the primary focus of this book is the mental health counseling realm, other related settings offer unique and valuable experiences for the counseling intern. These include school counseling, higher education counseling, and employee assistance counseling settings.

The school counselor intern, by virtue of her or his placement setting, is on the front lines of the counseling profession. He or she often confronts issues at the grassroots level, with home problems frequently becoming school problems. School counselors sometimes face parents' desperate, but unfair, demands to do intensive therapy or to handle children with severe psychological problems. Moreover, unfortunately, it has been our experience that some school counselors are asked to perform duties and assume roles that appear to directly conflict with counseling duties and roles. For instance, some school counselors are asked to become disciplinarian, assistant principal, test administrator (as sole function), or class scheduler. Very often these tasks incorporate an authoritarian or clerical role, which conflicts with training as a nonjudgmental and skilled professional counselor. Role conflict may result, causing consternation, role confusion, and stress.

Ideally, the school counselor does just what the title denotes: he or she counsels. Implicit in the role of a school counselor are various functions appropriate to his or her professional training, such as career, vocational, and college guidance and testing; counseling regarding situational concerns; referral to other professionals when appropriate (such as to the school psychologist or other mental health professional external to the school for problems that appear to be beyond the scope of the school counselor's role and training); parent conferences; teacher consultation and training in counseling areas; and so on. Serious psychological and psychiatric disorders should be treated in a clinical setting.

Higher education counseling can also be a rewarding experience for the counselor intern. Many placements are possible at the college or university level, including the counseling center, career and placement office, student development center, office of multicultural affairs, and other student-related departments such as housing and athletics. Because of the diversity of placement opportunities, work can vary according to the particular setting, from clinical mental health counseling at the counseling center to career and vocational counseling at the career services or placement center to performance enhancement in the athletic department.

Employee alcoholism programs of the 1940s and 1950s have been organized into employee assistance programs in more recent decades. Employee assistance counselors assist in identifying and resolving productivity problems associated with employees concerned with personal troubles such as health problems, marital or family issues, financial concerns, alcohol or drug problems, legal or emotional problems, and stress. Often employee assistance counselors provide consultation and training in job performance areas, referrals for diagnosis and treatment, linkages between workplace and community resources, and follow-up services for employees.

Whereas the early development of EAPs involved organized labor, management, and Alcoholics Anonymous, in more recent times EAPs have become more professional, with many counselors, social workers, and psychologists becoming certified in employee assistance. In addition, many states are moving to license individual EAP providers.

## A CHECKLIST FOR GETTING STARTED

To help you complete all the steps in selecting your internship site, we have compiled the following checklist:

1. Determine your own interests, needs, and expectations for the internship.
2. Review your counseling or human services program regulations for internships.
3. Ascertain any state requirements for supervised work experiences related to counseling licensure/certification.
4. Think about what kinds of educational opportunities you hope to experience during your internship, and what qualities you hope to find in your supervisor.
5. Make an appointment with your faculty adviser to discuss placement possibilities.
6. Make an appointment with your program placement coordinator to acquire further information.
7. Locate additional sources for information about potential internship sites, including other students, newspaper advertisements, and funding agencies.
8. Compose or update a resume and make several typed copies.
9. Arrange informational interviews as often as possible.

10. Follow up on each interview with a thank-you note and, a week or two later, a phone call.

11. Obtain and compare job descriptions for the internship sites that interest you.

12. Try to find a good fit for the placement, keeping in mind your own needs and interests, your program requirements, state licensing/certification regulations, and the unique experiences offered by and the individual characteristics of each internship site.

13. Secure a final agreement with the supervisor after you have selected a placement site.

14. Complete all official forms to register your supervised experience with the state licensing/certification board.

15. Determine the necessity of obtaining liability insurance. The American Counseling Association offers student members coverage at minimal cost (800-347-6647).

16. Ask your future supervisor how you can best prepare for your placement. Follow through on his or her suggestions.

17. Assess your academic record for areas relevant to the internship that you need to review.

18. Set up a schedule to read and review material so that you will begin your placement feeling as competent and comfortable as possible.

## BIBLIOGRAPHY

BRADLEY, F. (Ed.). (1991). *Credentialing in counseling.* Alexandria, VA: American Counseling Association.

COLLISON, B., & GARFIELD, N. (1990). *Careers in counseling and development.* Alexandria, VA: American Counseling Association.

COREY, G. (1996). *Manual for theory and practice of counseling and psychotherapy* (5th ed.). Pacific Grove, CA: Brooks/Cole.

COREY, G. (1996). *Theory and practice of counseling and psychotherapy* (5th ed.). Pacific Grove, CA: Brooks/Cole.

EMPLOYEE ASSISTANCE PROFESSIONAL ASSOCIATION (1992). *Standards for employee assistance programs.* Arlington, VA.

GOLDBERG, C. (1992). *The seasoned psychotherapist: Triumph over adversity.* New York: W.W. Norton.

HEPPNER, P. (Ed.). (1990). *Pioneers in counseling and development: Personal and professional perspectives.* Alexandria, VA: American Counseling Association.

HERR, E. (1989). *Counseling in a dynamic society: Opportunities and challenges.* Alexandria, VA: American Counseling Association.

HERRING, R. (1998). *Career counseling in schools: Multicultural and developmental perspectives.* Alexandria, VA: American Counseling Association.

JOHNSON, M., CAMPBELL, J., & MASTERS, M. (1992). Relationship between family of origin dynamics and a psychologist's theoretical orientation. *Professional Psychology: Research and Practice, 23*(2), 119–122.

KOMIVES, S., WOODARD, D., & DELWORTH, U. (Eds.). (1996).

*Student services: A handbook for the profession.* San Francisco: Jossey-Bass.

LEWIS, M., HAYES, R., & LEWIS, J. (1986). *The counseling profession.* Itasca, IL: Peacock.

MYERS, W. (1992). *Shrink dreams.* New York: Simon and Schuster.

PECK, M. (1978). *The road less traveled.* New York: Simon and Schuster.

PEDERSEN, P. & CAREY, J. (1994). *Multicultural counseling in schools: A practical handbook.* Boston: Allyn & Bacon.

SCHMIDT, J. (1996). *Counseling in schools: Essential services and comprehensive programs.* Boston: Allyn & Bacon.

SIEGEL, S., & LOWE, E. (1992). *The patient who cured his therapist.* New York: Dutton.

SKOVHOLT, T., & RONNESTAD, N. (1992). Themes in therapist and counselor development. *Journal of Counseling and Development, 70*(4), 505–515.

SUSSMAN, M. (1992*). A curious calling: Unconscious motivation for practicing psychotherapy.* Northvale, NJ: Jason Aronson.

TALLEY, J. & ROCKWELL, W. (1986). *Counseling and psychotherapy with college students.* Westport, CT: Praeger.

WALLACE, S., & LEWIS, M. (1990). *Becoming a professional counselor.* London: Sage.

YALOM, I. (1974). *Every day gets a little closer.* New York: Basic Books.

YALOM, I. (1989). *Love's executioner.* New York: Basic Books.

# 2

# Developing Competencies and Demonstrating Skills

Your internship provides an arena for you to try your wings as a helping professional with guidance and support close at hand. Many students feel a bit overwhelmed as they begin to interact with clients and staff members. You may have difficulty at first as you try to remember counseling theory and techniques, to recall academic coursework concerning such areas as human development or multicultural issues, to keep in mind ethical guidelines, and to think about agency procedures, regulations, and policies—all while trying to attend to your first few clients! More than one intern has felt discouraged after the first week of trying to juggle all the responsibilities of the new role.

Your internship can be viewed as a time to build a framework of new professional relational skills on a foundation of the material you have learned in your counseling program courses, your own life experiences, and your personal values and philosophies. This framework is composed of new perspectives, understandings, abilities, and skills added gradually and with care. Your goal is to construct a strong framework over a solid foundation, working diligently but patiently, and often standing back to look at the work you have accomplished so far.

During your internship you will develop some of the specific personal attributes and competencies that you will use during your professional counseling career. To help you delineate your goals, we have compiled the following list of skills for graduate-level internship students to work toward building. In reality, not all placement sites afford the opportunity to develop abilities in every area we have indicated. In addition, the quantity and scope of the competencies listed here reflect our belief that becoming a professional counselor is an ongoing process. Your internship is just the beginning of your professional development; you will continue to add competencies throughout your career.

## SUGGESTED COMPETENCIES FOR INTERNS

I. Communication Skills
  A. Verbal skills
    1. Students will be able to express themselves clearly and concisely in daily interactions with agency staff members and other professionals.
    2. Students will be able to communicate pertinent information about clients and to participate effectively in interdisciplinary treatment team meetings and case conferences, while maintaining their identities as counselors within a multidisciplinary group.
    3. Students will be able to educate clients and to provide appropriate information on a variety of issues (such as parenting, after-care and other support services, psychotropic medications, stress management, sexuality, or psychiatric disorders) in an easily understandable manner.
    4. Students will be able to communicate with clients' families, significant others, and designated friends in a helpful fashion. They will be able to provide, as well as to obtain, information concerning the client, while respecting the client's rights concerning privacy, confidentiality, and informed consent.
    5. Students will be able to communicate effectively with referral sources, both inside and outside the agency, concerning all aspects of client needs and well-being (for example, housing, legal issues, Twelve Step programs, and psychiatric concerns).
  B. Writing skills
    1. Students will be able to prepare a complete, written initial intake assessment, including a mental status evaluation, a psychosocial history, a diagnostic impression, and recommended treatment modalities.
    2. Students will be able to write progress notes, to chart, and to maintain client records according to agency standards and regulations.
    3. Students will be able to prepare a written treatment plan, including client problems, therapeutic goals, and specific interventions to be utilized.
    4. Students will be able to prepare a formal, written case study.
    5. Students will be able to use computer skills to work with word-processing programs and to maintain and search databases.

    C. Knowledge of nomenclature
1. Students will thoroughly know professional terminology pertaining to counseling, psychopathology, treatment modalities, and psychotropic medications.
2. Students will be able to understand professional counseling jargon and will be able to participate in professional dialogue.

II. Interviewing
    A. Students will structure the interview according to a specific theoretical perspective (for example, psychodynamic or behavioral theory) because a theory base provides the framework and rationale for all therapeutic strategies, techniques, and interventions.
    B. Students will be able to use appropriate counseling techniques to engage the client in the interviewing process, to build and maintain rapport, and to begin to establish a therapeutic alliance. This may include using attending behaviors, active listening skills, and a knowledgeable and professional attitude to convey empathy, genuineness, respect, and caring, and to be perceived as trustworthy, competent, helpful, and expert (Lewis, Hayes, & Lewis, 1986).
    C. Students will be able to use appropriate counseling techniques to increase client comfort and to facilitate collection of data necessary for clinical assessment, such as evaluating mental status, taking a thorough psychosocial history, and eliciting relevant, valid information concerning the presenting problem, in order to formulate a diagnostic impression. Specific interviewing competencies may include observation, use of open-ended and closed-ended questions, the ability to help the client stay focused, reflection of content and feeling, reassuring and supportive interventions, and the ability to convey an accepting and nonjudgmental attitude.
    D. Students will develop a holistic approach toward interviewing by assessing psychological, biological, environmental, and interpersonal factors that may have contributed to the client's developmental history and presenting problems.
    E. Students will strive to see things from the client's frame of reference and to develop a growing understanding of the client's phenomenological perspective.
    F. Students will be aware at all times of the crucial importance of understanding the client from a multicultural perspective and will be aware that sociocultural heritage is a key factor in determining the client's unique sense of self, worldview, values, ideals, patterns of interpersonal communication, family structure, behavioral norms, and concepts of wellness as well as of pathology.

III. Diagnosis
    A. Students will understand the most commonly used assessment instruments, such as personality and intelligence tests, anxiety and depression scales, and interest inventories.

1. Students will become familiar with the validity and reliability of these instruments.
2. Students will be able to interpret data generated by these instruments and understand the significance of these data in relation to diagnosis and treatment.
3. Students will be able to determine which assessment instruments would be most helpful in evaluating specific client problems or concerns.
4. Students will be aware of the limitations of assessment instruments when used with ethnic minority populations.

B. Students will develop a working knowledge of the DSM-IV.
1. Students will be familiar with the organization of the DSM-IV and will be able to use this nosology effectively (for example, to find diagnostic codes or to trace clients' behaviors, affects, or cognitions along the decision trees to ascertain potential diagnoses).
2. Students will be able to understand the DSM-IV classification of disorders and will be able to identify particular constellations of client problems as specific DSM-IV diagnostic categories.

C. Students will be able to review and consider all pertinent data, including interviews, medical records, previous psychiatric records, test results, psychosocial history, consultations, and DSM-IV classifications, in formulating a diagnostic impression or preliminary diagnosis.

IV. Treatment
A. Students will be able to conduct therapy using accepted and appropriate treatment modalities and counseling techniques based on recognized theoretical orientations.
1. Students will work toward identifying their own theoretical frameworks based on their own philosophy of humankind.
2. Students will know how to make treatment recommendations, formulate a treatment plan, establish a treatment contract, implement therapy, and terminate the therapeutic relationship at an appropriate time. Refer to the sample treatment plan form in Appendix I.
3. Students will be able to conduct the following types of therapy and will understand the underlying principles, issues, dynamics, and role of the counselor associated with each type of treatment:
   a. Conjoint therapy
   b. Crisis intervention
   c. Family therapy
   d. Group therapy
   e. Individual therapy
   f. Marital therapy

B. Students will understand that different client populations and different types of problems may respond best to varying therapeutic approaches and techniques.

1. Students will be knowledgeable about various types of client populations and their particular problems and concerns, including but not limited to the following:
   a. Adult children of alcoholics
   b. Adults
   c. Chemically dependent individuals
   d. Children and adolescents
   e. Clients of varied ethnic and religious backgrounds
   f. Dual-diagnosed clients (for example, chemically dependent with a psychiatric disorder)
   g. Individuals with eating disorders
   h. Elders
   i. Gay and lesbian clients
   j. Physically or cognitively impaired clients
   k. Survivors of trauma and abuse
   l. Any other individual who may be included under the Americans with Disabilities Act
2. Students will be flexible and knowledgeable in determining population-appropriate counseling techniques and therapeutic interventions. Students will have as many therapeutic tools available for use as possible (for example, play therapy, art therapy, behavioral techniques, role-playing, gestalt techniques, directive versus nondirective techniques, stress management techniques, experiential therapy, hypnosis).

C. Students will be sensitive to the impact of multicultural issues on the counseling relationship and on treatment and will modify therapeutic approaches and techniques to respect multicultural differences and to meet multicultural needs.

V. Case Management

A. Students will understand the functions and goals of all departments, programs, and services within the agency and will be able to network with appropriate personnel throughout the social service system.

B. Students will understand the roles, responsibilities, and contributions to client care of members in each department or program within the agency. The student will know which individual(s) to contact to help resolve various client problems.

C. Students will acquire a thorough knowledge of community resources and will understand the agency procedures for referring clients to outside sources for help.

D. Students will consider continuity of care to be a most important goal, beginning with the initial client contact.

   1. Students will act as an advocate for the client in ensuring continued quality of care and access to social services. Advocacy will include, but not be limited to, exploring possible funding sources for care, such as mental health coverage on insurance policies, Medicaid, or Medicare.

2. Students will be able to participate in all areas of discharge planning, including arranging follow-up visits with a mental health professional, communicating with insurance companies, and providing help with housing, transportation, vocational guidance, legal assistance, support groups, medical care, and referral to other services or agencies.

VI. Agency Operations and Administration

A. Students will be familiar with the organizational structure of the agency and will understand the responsibilities and functions of administrative staff.

B. Students will understand the philosophy, mission, and goals of the agency and will thoroughly know the agency's policies and procedures, which are usually delineated in a comprehensive manual.

C. Students should be aware of immediate and long-range strategic plans for the agency (for example, to hire an art therapist, to develop a chemical abuse program, or to add an additional building, as well as to evaluate and eliminate ineffective programs).

D. Students will understand the business aspects of the agency (for example, funding sources or budget allowances).

E. Students will be aware of legal issues concerning agency functions, such as state or national licensure/certification requirements or safety regulations.

F. Students will understand agency standards that ensure continued quality of care, including quality assurance and peer review processes.

VII. Professional Orientation

A. Students will know all ethical and legal codes for counselors, provided by professional counseling associations as well as by state law, and will adhere to these standards at all times.

B. Students will be familiar with agency regulations and policies regarding ethical and legal issues and will adhere to these standards at the placement site.

C. Students will be knowledgeable concerning legislation protecting human rights.

D. Students will seek guidance from the on-site supervisor and the academic program supervisor with any questions concerning ethical or legal issues or professional behavior.

E. Students will consider the four basic Rs for counselors (Carkhuff, 1987) whenever acting in a professional helping capacity: the right of the counselor to intervene in the client's life, the responsibility the counselor assumes when intervening, the special role the counselor plays in the helping process, and the realization of the counselor's own resources in being helpful to the client.

You may wish to look at the sample evaluation of intern form in Appendix K as well as Appendix L's sample intern evaluation of site and supervisor form.

# REFERENCES

LEWIS, M., HAYES, R., & LEWIS, J. (1986). *The counseling profession.* Itasca, IL: Peacock.

CARKHUFF, R. (1987). *The art of helping* (Vol. VI). Amherst, MA: Human Resource Development Press.

# BIBLIOGRAPHY

AMERICAN PSYCHIATRIC ASSOCIATION. (1994). *Diagnostic and statistical manual of mental disorders* (4th ed.). Washington, DC.

ANDERSON, D. (1992). A case for standards of counseling practice. *Journal of Counseling and Development, 71*(1), 22–26.

AXLINE, V. (1969). *Play therapy.* New York: Random House.

BALSAM, M., & BALSAM, A. (1984). *Becoming a psychotherapist* (2nd ed.). Chicago: University of Chicago Press.

BENJAMIN, A. (1987). *The helping interview.* Boston: Houghton Mifflin.

BURN, D. (1992). Ethical implications in cross-cultural counseling and training. *Journal of Counseling and Development, 70*(5), 578–583.

CASEMENT, P. (1991). *Learning from the patient.* New York: Guilford.

CHAPLIN, J. (1985). *Dictionary of psychology* (2nd ed. rev.). New York: Doubleday.

CHESSICK, R. (1989). *The technique and practice of listening in intensive psychotherapy.* New York: Jason Aronson.

CHESSICK, R. (1993). *A dictionary for psychotherapists: Dynamic concepts in psychotherapy.* Norwalk, NJ: Jason Aronson.

ELKIND, S. (1992). *Resolving impasses in therapeutic relationships.* New York: Guilford .

ENGELS, D., & DAMERON, J. (Eds.). (1990). *The professional counselor: Competencies, performance, guidelines, and assessment* (2nd ed.). Alexandria, VA: American Counseling Association.

FAIVER, C. (1992). Intake as process.

*Journal of the California Association for Counseling and Development, 12,* 83–86.

FAIVER, C., & O'BRIEN, E. (1993). Assessment of religious beliefs form. *Counseling and Values, 37*(3), 176–178.

FALVEY, J. (1992). From intake to intervention: Interdisciplinary perspectives on mental health treatment planning. *Journal of Mental Health Counseling, 14*(4), 471–489.

HOOD, A., & JOHNSON, R. (1991). *Assessment in counseling: A guide to the use of psychological assessment procedures.* Alexandria, VA: American Counseling Association .

HUGHES, J., & BAKER, D. (1990). *The clinical child interview.* New York: Guilford.

KARASU, T. (1992). *Wisdom in the practice of psychotherapy.* New York: Basic Books.

KOTTLER, J. (1992). *Compassionate therapy: Working with difficult clients.* San Francisco: Jossey-Bass.

KURPIUS, D., & BROWN, D. (1988). *Handbook of consultation: An intervention for advocacy and outreach.* Alexandria, VA: American Counseling Association.

L'ABATE, L., FARRAR, J., & SERRITELLA, D. (Eds.). (1992). *Handbook of differential treatments for addictions.* Boston: Allyn & Bacon.

LEE, C., & RICHARDSON, B. (1991). *Multicultural issues in counseling.* Alexandria, VA: American Counseling Association.

MITCHELL, R. (1991). *Documentation in counseling records* (Vol. 2). Alexandria, VA: American Counseling Association.

MORROW, K., & DEIDAM, C. (1992). Bias in the counseling profession: How to recognize and avoid it. *Journal of Counseling and Development, 70*(5), 571–577.

MYERS, J. (Ed.). (1994). *Developing and directing counselor education laboratories.* Alexandria, VA: American Counseling Association.

OGDON, D. (1990). *Psychodiagnostics and personality assessment: A handbook* (2nd ed.). Los Angeles: Western Psychological Services.

PARAD, H., & PARAD, L. (1990). *Crisis intervention* (Vol. 2). Milwaukee: Family Service America.

PEDERSON, P. (1988). *A handbook for developing multicultural awareness.* Alexandria, VA: American Counseling Association.

REID, W., & WISE, M. (1989). *DSM-III-R training guide.* New York: Brunner/Mazel.

SHAPIRO, D. (1965). *Neurotic styles.* New York: Basic Books.

SHEA, S. (1988). *Psychiatric interviewing: The art of understanding.* Philadelphia: Saunders.

SPITZER, R., GIBBON, M., SKODOL, A., WILLIAMS, J., & FIRST, M. (1989). *DSM-III-R case book.* Washington, DC: American Psychiatric Press.

STEENBARGER, B. (1991). All the world is not a stage: Emerging contextualist themes in counseling and development. *Journal of Counseling and Development, 70*(2), 288–296.

STREAN, H. (1993). *Resolving counter-resistances in psychotherapy.* New York: Brunner/Mazel.

SWARTZ, S. (1992). Sources of misunderstandings in interviews with psychiatric patients. *Professional Psychology: Research and Practice, 23*(1), 24–29.

WALLACE, S., & LEWIS, M. (1990). *Becoming a professional counselor.* London: Sage.

YALOM, I. (1985). *Theory and practice of group psychotherapy.* New York: Basic Books.

# 3

# The Site Supervisor

## GENERAL RESPONSIBILITIES

The internship site supervisor has a highly significant role and many responsibilities for your training. With this individual, you will begin to demonstrate the knowledge and skills you have acquired through your formal academic training. The professional and personal relationship you establish with your site supervisor will set the pace, direction, and tone for your internship. His or her willingness to supervise you becomes a commitment to assist you in attaining and maintaining a counseling relationship with clients.

Ronnestad and Skovholt (1993) note that beginning students value a supervisor who teaches and provides structure and direction to them. Other researchers emphasize the importance of supervisor support of interns, which may include assistance with client selection and provision of relevant agency and client information (a guidance function) (Grater, 1985); establishment of a milieu that creates a sense of security for the intern (Rabinowitz, Heppner, & Roehlke, 1986); and the provision of encouragement and feedback about progress (Worthington & Roehlke, 1979).

In many ways what follows is our conception of the ideal supervisor, the super supervisor. Let's call this person the SUPERvisor. This is certainly unfair to supervisors in that it is not always possible to demonstrate perfection in deed or in person. Nonetheless, for our purposes, we shall demand perfection!

You must feel professionally and personally comfortable with your supervisor and believe that this person will be a good role model for you. The time you spend together should provide ample opportunities for you to get to know each other as well as to assess your ability to work together. Even though the supervisor maintains major responsibility for the integrity of the internship and for encouraging you as you pursue a career in counseling, your belief that he or she can work with you is vital. Ideally, the supervisor and you share the hope that the internship will be a worthwhile and enjoyable experience for both of you. Any hesitancy you have should be discussed candidly with your site supervisor or university supervisor so any necessary accommodations can be made.

The internship experience is in many ways similar to the counseling relationship; therefore, you will want to work with a supervisor whose therapeutic orientation, ideology, and style are similar to yours.

One of the key functions of your supervisor is to provide feedback on your performance and progress. This critical feedback process will help you stay focused on both the quality and the quantity of your services. As part of the ACA Code of Ethics (see Appendix E, section F), your supervisor accepts the responsibility to help ensure that you develop into a competent counselor. By providing feedback and direction he or she provides a great preparatory service in your professional development.

A supervisor is an educator, and as such he or she enhances your formal academic training. The site supervisor's role modeling may uniquely influence your approach to the profession. Our ideal supervisor assists you in feeling comfortable in awkward or new situations. Moreover, we hope that your supervisor reveals humanity as well as his or her professional persona, and feels comfortable enough to let you know the intricacies of the profession by showing you his or her honest attitudes, emotions, and behavior. You will find his or her support and direction key factors in the refinement of your continuing journey into the counseling profession. The site supervisor's task is serious; he or she has made a commitment to counseling itself by agreeing to supervise you.

At times you may find that the site supervisor will share facets of the profession that you have not learned in textbooks. We anticipate that his or her experience, training, and skills will afford you the opportunity to learn much about the field. In essence, he or she might share "the good, the bad, and the ugly" of the counseling field, all of which can be valuable as you continue your professional development. Take all experiences as learning experiences; even the bad can be looked on (or as some say, reframed) as a challenge and an opportunity.

One issue students often address is how to interact with other mental health professionals at the site. Most likely, your supervisor will ask his or her colleagues to expose you to their professional styles, approaches, and theories. Through such exposure, you may find that their vantage point sometimes differs from your supervisor's. This can pose a dilemma and perhaps be confusing to you. Our ideal supervisor is acutely aware that his or her orientation is not absolute. However, he or she will likely desire that you adhere to his or her directives because ultimate responsibility rests with the supervisor. He or she will be willing to discuss with you any differences you have observed and, generally, will use such experiences as teaching tools. Ultimately, by internal (sometimes unconscious) reflection you will decide how all these bits of information, observations, and various approaches fit you as a therapist. In the interim, you will save yourself much frustration and cognitive dissonance by relying on this supervisor to achieve your educational goals and objectives. Work closely with your supervisor to make the internship experience an enjoyable, educational, and personally rewarding opportunity. Do not hesitate to share with him or her where your interests lie and what you like and dislike about the total experience, including agency ambience, clients, and staff. Being candid with your supervisor can optimize this experience, enhancing both your learning and your relationship. Your placement is your testing ground for academic knowledge and should be as pleasant and beneficial as possible for the best learning results. No doubt your supervisor will share this goal with you.

As noted previously, this is probably the final stage of your academic endeavor and, as such, is the pinnacle of a course of study. Among the many thoughts that might occur to you at this point is "Do I really want to be a counselor?" Don't be concerned about such thoughts; it is our experience that they may reflect self-doubt. Be willing to explore thoughts such as these with your supervisor. Also, integrating your formal education with practice may cause you to realize that the counseling field is not exactly what you thought, were told, or read it would be. Believing something about the profession without validating it does a great disservice to yourself, your colleagues, and, above all, your future clients. Use the placement to solidify your beliefs and attitudes about the profession. Because your site supervisor and university academic adviser have worked with you up to this point to establish your professional goals and objectives, it is most appropriate to discuss this key factor as well.

Often separating your professional and personal involvement with your supervisor is difficult because the nature of the placement can cause blurred role boundaries. Your supervisor adheres closely to the ACA Code of Ethics and Standards of Practice (see Appendix E), so his or her personal side can influence the tone, atmosphere, and ethos of the placement. Exposing his or her shortcomings as well as strengths reinforces the uniqueness and value of your relationship. Further, being a person first enhances the value of the quality of interaction.

It is a rare supervisor who does not learn from his or her students. For example, your current involvement in the academic arena may augment his or her knowledge base. Your supervisor may take pride in helping to shape and mold your career, certainly a powerful personal learning experience. The relationship you share can leave a positive and lasting impression on you both.

## BEFORE INTERNSHIP

Now let's look at this special relationship as it develops from beginning to end. A number of tasks need to be achieved when you first meet your supervisor. He or she will need to know your university and state licensing or certification requirements as well as your personal goals. The placement can influence your belief about the counseling profession itself, so it may be helpful for the supervisor to require definitive, objective, and measurable goals. Additionally, he or she has the responsibility to understand who you are, especially as an aspiring counseling professional. Because the supervisor is aware of the placement site's policies and procedures, he or she will gauge the appropriateness of a match and the potential for successful completion. It is not unusual for the supervisor to conduct a somewhat formal interview with you to learn why you chose this site, what your career objectives are, and what you would like to achieve in your professional relationship with him or her. Your level of understanding regarding theories of counseling, techniques, and the counseling field in general will also be important. The supervisor will assess whether you will be able to integrate your book knowledge from your academic training at this specific site. In addition to examining your academic foundation, your supervisor will also consider your personal attributes.

No two training sites have the same mix of personnel, client population, and agency standards and ethos. For these reasons, the supervisor, as presumed expert of his or her professional domain, must make the final decision regarding your acceptance. Once you are accepted, however, our ideal supervisor will outline for you any forms you need to complete before starting the placement. The supervisor may also suggest specific readings, including pertinent agency orientation materials such as policies and procedures. Perhaps you will be asked to review your theories and techniques as well as to start solidifying a therapeutic orientation based on your knowledge of the clinical population. You may also be asked to prepare audiotape or videotape role-play demonstrations so the supervisor can establish a baseline of your knowledge and skill level. The supervisor may discuss with you how you feel personally about working with certain clinical populations to see if any personal issues may impede your ability to be effective and efficient with clients. Professional involvement will not be devoid of the personal feelings, attitudes, and beliefs that are a part of who you are.

As the final preparations are made for your placement, the supervisor will explain in depth the agency's policies, procedures, mission statements, and other information relevant to your placement. Specific attention may be given to past involvement of students at this site and how they fit into the overall functioning of the organization. The supervisor will help you to understand how you will be perceived during your placement, what specific parameters will be in place, and how you will fit into the organization. Your supervisor will be an advocate for you and has agreed to shoulder the responsibility to have you integrate successfully into the treatment milieu and the organizational structure. The routine and seemingly mundane tasks of completing the necessary academic course forms (including registration) and student/supervisor agreements should be done at your earliest convenience. Watch for university and agency deadlines! If you are governed by specific licensure and/or certification requirements, understanding these standards and promptly filling out any necessary forms are vital to receiving proper experiential credit.

## DURING INTERNSHIP

Because internship is a great opportunity to put your skills and knowledge to use, our exemplary supervisor will provide you with the chance to use them to the fullest. The ideal is to have you experience all the clinical aspects possible within the organization. What better way to delve into the "real world" of counseling than to practice one's skills in all areas of clinical services?

Barring any sensitive issues for which clients may prefer not to have you present, it is appropriate and a valuable learning experience to "shadow" the supervisor in all of his or her clinical activities. By observing the supervisor and eventually participating in services, you will gradually become acclimated to the setting and begin integrating your book knowledge with clinical practice. Shadowing lets both of you assess the scope of your knowledge and level of comfort as you aspire to independent work with clients. When you accept the responsibility of independent client contact depends on a consensus between you and your supervisor as to your readiness. Further, our clients unfortunately do not always fit into our plans as to when and what types of techniques or approaches we would like to practice with them. You should feel comfortable discussing with your supervisor your perceived readiness to accept more responsibility and to use your knowledge and skills. Take note of how it feels to use your knowledge.

One critical issue to address throughout the internship is what treatment approach to use with your clients. Before you solidify your approach, review with your supervisor your philosophy of humankind (see Chapter 9 on self-assessment). Perhaps you have previously dealt with this personal philosophical stance in your theoretical studies, but now you will link those theories with your belief system in a practical setting with contact with fellow human beings, who are, for the most part, suffering. We believe that your philosophy of humankind affects the way you approach and treat your clients, and, for that matter, people in general. For many students this will be the first time they closely examine how their beliefs can influence their dealings with clients in vivo. Once you begin to formulate, review, and refine your philosophy of humankind, our ideal supervisor can guide you in selecting a philosophically and theoretically congruent treatment approach. For example, existential approaches lend themselves more to freedom of choice, whereas analytically and behaviorally oriented approaches appear to be deterministic.

Throughout the internship your supervisor will schedule both formal and informal supervision conferences with you. During these meetings you will review your internship goals and objectives to assess adherence, progress, and any necessary modifications to them. Your supervisor will give you constructive feedback as to the quality of your performance, including recommendations for improvement. You will generally be formally evaluated midway through the placement and again at the end; your academic adviser may be part of this process as well. Particular attention may be given to how well you feel you have become part of the agency's treatment structure and the overall treatment team;

your supervisor will validate your perceptions of these two areas. In addition, your supervisor may help you consider whether a specific clinical population or treatment specialty may be indicated and elaborate on any areas in which you are demonstrating competency (see Chapter 2, which discusses competencies you should develop).

## AT THE TERMINATION OF INTERNSHIP

Your site supervisor has served in a highly significant role and maintained a considerable amount of responsibility throughout your internship. One of his or her most enjoyable tasks is to help you bring closure to your placement experience. Your supervisor has worked closely with you by investing time, knowledge, and skill to help you become a competent counselor-in-training. He or she will be eager to share impressions and recommendations with you as to your clinical competencies, skills, and knowledge base. The supervisor will provide feedback on how well you were able to translate your academic foundation to the practical work situation. His or her input, guidance, and direction can influence your career focus. This is a responsibility in which the supervisor takes pride. The completion of your final evaluation, often in concert with your academic adviser, will formally appraise your efforts. This formal document presents measurable, objective, and observable data about your clinical strengths and weaknesses, general impressions, and overall recommendations. You should have ample opportunity to respond to the evaluation and, if necessary, get clarification on its content. Further, you may be invited to evaluate both your site and supervisor in what we view as a reciprocal process. (Appendices K and L present sample evaluation forms of both the intern and the site and supervisor.)

## A WORD ABOUT YOUR
## UNIVERSITY SUPERVISOR

We assume that by the time you get to the internship placement you have formed a relationship with your university supervisor. Perhaps you have had him or her for classes previously; perhaps the university supervisor serves as your program adviser. In any case, under most circumstances this person serves as the university liaison to your agency and professor of the integrating internship course on campus. Most programs offer a weekly or biweekly seminar on campus during which interns process their experiences and present and critique cases (of course masking any identifying client data). We strongly suggest that you take any and all placement problems to your university supervisor early so as to avoid frustration and consternation.

# REFERENCES

GRATER, H. (1985). Stages in psycho-therapy supervision: From therapy skills to skilled therapist. *Professional Psychology: Research and Practice, 16,* 605–610.

RABINOWITZ, F., HEPPNER, P., & ROEHLKE, H. (1986). Descriptive study of process and outcome variables of supervision outcome. *Journal of Counseling Psychology, 33*(3), 292–300.

RONNESTAD, M., & SKOVHOLT, T. (1993). Supervision of beginning and advanced graduate students of counsel-ing and psychotherapy. *Journal of Coun-seling and Development, 71*(4), 396–405.

WORTHINGTON, E., & ROEHLKE, H. (1979). Effective supervision as per-ceived by beginning counselors in training. *Journal of Counseling and Devel-opment, 26*(1), 64–73.

# BIBLIOGRAPHY

ALLEN, G., SZOLLOS, S., & WILLIAMS, B. (1986). Doctoral students, compara-tive evaluations of best and worst psychotherapy supervision. *Professional Psychology: Research and Practice, 17,* 91–99.

BARTLETT, W., GOODYEAR, R., & BRADLEY, F. (Eds.). (1983). Supervi-sion in counseling II (Special issue). *The Counseling Psychologist, 11*(1).

BORDERS, L., & LEDDICK, G. (1987). *Handbook of counseling supervision.* Alex-andria, VA: American Counseling Association.

CHIAFERI, R., & GRIFFIN, M. (1997). *Developing fieldwork skills: A guide for human services, counseling, and social work students.* Pacific Grove, CA: Brooks/Cole.

ELLIS, M. (1991). Critical incidents in clinical supervision and in supervisor supervision: Assessing supervisory issues. *Journal of Counseling Psychology, 38*(3), 342–349.

GLIDDEN, C., & TRACEY, T. (1992). A multidimensional scaling analysis of supervisory dimensions and their per-ceived relevance across trainee experi-ence levels. *Professional Psychology: Re-search and Practice, 23*(2), 151–157.

HESS, A. (1980). *Psychotherapy supervision: Theory, research and practice.* New York: Wiley.

KAISER, T. (1997). *Supervisory relationships: Exploring the human element.* Pacific Grove, CA: Brooks/Cole.

LEWIS, M., HAYES, R., & LEWIS, J. (1986). *An introduction to the counseling profession.* Itasca, IL: Peacock.

MASTERS, M. (1992). The use of positive reframing in the context of supervi-sion. *Journal of Counseling and Develop-ment, 70*(3), 387–390.

MUNSON, C. (Ed.). (1984). *Supervising student internships in human services.* Binghamton, NY: Haworth.

PITTS, J. (1992). Pips: A problem-solving model for practicum and internship. *Counselor Education and Supervision, 32*(2), 142–150.

SKOVHOLT, T., & RONNESTAD, M. (1992). Themes in therapist and coun-selor development. *Journal of Counseling and Development, 70*(4), 505–515.

SWANSON, J., & O'SABEN, C. (1993). Differences in supervisory needs and expectations by trainee experience, cognitive style, and program member-ship. *Journal of Counseling and Develop-ment, 71*(4), 457–464.

VASQUEZ, M. (1992). Psychologist as clinical supervisor: Promoting ethical practice. *Professional Psychology: Research and Practice, 23*(3), 196–202.

YERUSHALMI, H. (1992). On the con-cealment of the interpersonal thera-peutic reality in the course of supervi-sion. *Psychotherapy, 29*(3), 438–446.

# 4

# Deciding How to Help: Choosing Treatment Modalities

By the time you are ready to begin your internship, you will have completed all or most of the academic courses required for your degree in counseling. You will have studied a variety of theoretical models and clinical techniques, and you will have acquired substantial knowledge of many other academic areas that are fundamental to counseling. This chapter provides a practical framework to help you synthesize all this scholarly information and formulate a rational protocol for selecting treatment modalities. In addition, we highlight the theoretical concepts and specific counseling methods we feel are most salient for you as a counselor intern.

## THE COUNSELING PROCESS

Counseling may be conceptualized as an interactive process involving a trained professional counselor (or counselor intern) and a client, with the purpose of enhancing the client's level of functioning. Counselor and client work together as allies, helping the client grow and change by identifying goals, developing new ways of understanding and managing problems, and learning to use internal and environmental resources more effectively. However, as a counselor intern, you may wonder just how to begin helping the client who presents multiple, complicated problems, has no social support system, and says, "I'm a hopeless case. I've already tried everything and nothing helps."

When working with clients during your internship, it is helpful to remember that the counseling process always involves the following stages, regardless of your theoretical orientation, your level of clinical experience, or the complexity of the client's problems:

- Relationship building (developing trust and building a therapeutic alliance)

- Exploration (developing a full understanding of the client and his or her problems)

- Decision making (deciding on specific goals and strategies to achieve them)

- Implementation (using internal and external resources to resolve concerns)

- Termination and follow-up (assessing readiness to end therapy and ensuring that progress is maintained) (Doyle, 1998)

These stages are interrelated and interlocking rather than linear or sequential. Clients may need to move through some of them more than one time. For example, difficulties with implementation may lead to further work on decision making or exploration.


## ESTABLISHING TRUST

As a counselor intern, you can begin to help by establishing rapport with your client and working to build trust. The development of trust is the key factor in counseling; trust is the underlying force that enables the counselor and client to negotiate the other stages of the counseling process (Velleman, 1992). However, trust is not a fixed entity that may be taken for granted once it has been achieved. Trust is tested by the client and may need to be reestablished by the counselor many times, depending on the client's personality, developmental history, and current life situation. Often, each time the client faces a challenging new issue or a painful or frightening feeling, trust is questioned once again.

When you select a treatment modality, you will need to consider the current level of trust your client feels; some therapeutic interventions may be too uncomfortable for clients who feel unsafe. Therefore, working to help your client feel safe is crucial. Always be reliable and dependable. Prove your trustworthiness in basic, concrete ways. We have found the following behaviors to be especially helpful in developing client trust:

- Start and end sessions on time, and do your best to maintain the regularity of the appointment schedule that you and your client have agreed upon.

- Make certain that your client has your full attention and respect during sessions, and try to demonstrate your caring in both verbal and nonverbal ways.

- Always follow through on counseling-related tasks, such as finding a support group or a list of helpful books for your client.

- Always maintain strict confidentiality; never discuss a client outside the treatment setting unless you first have a signed release and have explained to your client exactly what information will be disclosed or exchanged.

- Do not talk about specific details of problems of other clients, even without mentioning their names. Clients may feel that their own time with the counselor is intruded upon when attention is focused on other clients during the session. Clients may also worry about a breach of confidentiality if the counselor discloses certain aspects of other clients' problems.

- Maintain professional boundaries at all times: Do not touch your client physically; do not engage in any social interactions outside the counseling relationship; do not disclose intimate personal information or discuss your own problems; do not participate in any business transactions with your client except for specified agency arrangements for the client to pay for services (Gutheil & Gabbard, 1993).

## THERAPEUTIC TOOLS

During your internship you will be developing a repertoire of therapeutic tools to help your client through each of the counseling stages just mentioned. These tools include your personal style of engaging a client, your ability to attend to both verbal and nonverbal cues in order to understand the client fully, and your specific counseling methods and interventions. The tools you choose should feel comfortable to you and should be appropriate for the particular client, the client's unique concerns, and the current stage of the counseling process.

A wide range of therapeutic tools and flexibility in selecting them help ensure that you will have plentiful resources to assist your clients during your internship. We are not advocating a random assortment of counseling techniques or an undisciplined eclecticism with no regard for the principles of the theoretical schools. We suggest, instead, that your choice of techniques be grounded in your knowledge of counseling theory while it is guided by:

- The individual needs and goals of the client.

- The immediate problems of the client.

- The present stage of the counseling process.

- Your evolving relationship with the client.

- Your personal philosophy of life and your own understanding of humankind.

Corey (1991a) recommends an integrative perspective for the student counselor:

**More Counselor Activity**

Behavioral therapies

Cognitive/behavioral therapies

Gestalt therapy

Existential therapies

Client-centered therapy

Psychodynamic therapies

Classic psychoanalytic therapy

**Less Counselor Activity**

**FIGURE 4.1** Counselor Activity Level in Varying Therapeutic Orientations

Comparing your own thinking on the underlying issues with the positions of the various therapies can assist you in developing a frame of reference for your personal style of counseling (p. 178).

Your personal counseling style is in part determined by your activity level in counseling, which includes the frequency of your interventions and your role in controlling the direction of the session/therapeutic process. The various therapies may also be arranged along a continuum of counselor activity level, with classic psychoanalytic schools on the less active end and behavioral schools on the more active end (see Figure 4.1). You may wish to compare your own comfort level and personal style with the counselor activity levels of the theoretical schools.

## THE THERAPEUTIC RELATIONSHIP

Virginia Satir (1987) writes that although many diverse theories and techniques of therapy have been developed, "the basic ingredient remains the relationship between the therapist and the patient" (p. v). The primary importance of the helping relationship itself as a healing factor and as a pathway for client change, regardless of the counselor's theoretical orientation, has been noted by many theorists and clinicians (Carkhuff, 1984; Kurpius, 1986; Rogers, 1951, 1957, 1961). Counselor empathy has been recognized as the most central dynamic in determining the quality of the therapeutic relationship (Gilbert, 1992; Kahn, 1991; Kurpius, 1986; Shea, 1988). During your internship, your ability to be empathic, to "be there" and understand your client's internal experiences, will be healing for your client. As a novice counselor, you will be learning to use the professional relationship between yourself and your client to help your client move forward. This therapeutic relationship may be characterized or expressed through your selection of treatment approaches or modalities (Satir, 1987).

## UNDERSTANDING AND HELPING

### A Holistic Perspective

The first step in choosing an appropriate treatment modality or therapeutic style involves working to understand your client from a biopsychosocial perspective. We encourage counselor interns to take note of genetic, biological, and intrapsychic forces in relation to external influences of family, society, and world (Dacey & Travers, 1991). There is always a dynamic interplay between the client and the situation. As a counselor intern, it is important for you to use a comprehensive perspective, considering the client as an individual interacting with a complex, multidimensional environment.

Consider your client's problems, behaviors, and concerns in context at all times, and keep in mind that a seemingly abnormal behavior may be a normal response to an abnormal or highly stressful situation. Your client's problematic symptom can often be reframed as his or her solution to coping with internal or external stressors. Although the solution may have made sense at one time, it no longer works in the client's best interests. Together, you and your client can begin to understand the etiology and the purpose of the symptom in order to replace it with a more acceptable coping skill.

### A Developmental Perspective

During your internship, you will also find it useful to assess your client's personality traits, concerns, and behaviors within a framework provided by personality theorists and developmental psychologists. Clients often require help in negotiating critical turning points in the life span or in dealing with difficult developmental tasks (Corey, 1991b). Knowledge of developmental stages allows you to understand what types of problems are likely to arise at each specific phase of your client's life cycle so that you can be most helpful. Awareness of developmental norms enables you to help your client back onto a healthy developmental pathway.

In addition, your thorough knowledge of developmental stages provides a template for determining the relative health or pathology of your client

(Gabbard, 1990). Behavior that is appropriate and acceptable at one stage of life may be troublesome at another stage. For example, a young adolescent who becomes highly anxious and flees when faced with the prospect of going out on a date presents a vastly different clinical picture than does a 30-year-old individual with the same problem. As counselors, we are not concerned when a 2-year-old expresses disagreement by having a temper tantrum, but we see a red flag if a 25-year-old engages in such behavior. Similarly, we are not worried about the 80-year-old who spends hours reminiscing about the past, although we may be concerned about the 50-year-old who chooses to sit for hours talking about the past.

## A Multicultural Perspective

Ethnicity implies a set of prescribed values, norms, traits, communication styles, and behavioral patterns that constitutes a unique cultural reality and provides a sense of identity for members of the ethnic group (Lee, 1991). This ethnic identity colors the way the individual perceives the self and the world and provides meaning and structure to everyday activities, interpersonal relationships, and social behaviors (Gibbs & Huang, 1989). Ethnicity also determines the way people experience, express, and cope with distress, as well as how, where, and from whom they seek help (Root, 1985). Vontress (1986) points out that both clients and counselors are products of their cultures. Cultural factors therefore influence every aspect of the counseling process. Familiarity with multicultural issues will enable you to establish rapport and build the therapeutic relationship that is crucial for successful counseling.

Cultural sensitivity in counseling requires the knowledge base to recognize and balance universal norms, ethnic norms, and individual norms in assessing client concerns and selecting treatment modalities (Lopez, Grover, Holland, Johnson, Kain, Kanel, Mellins, & Rhyne, 1989). Cultural sensitivity helps you determine whether client behaviors, symptoms, or attitudes are actually ethnically based rather than defensive, inappropriate, or pathological. Being aware of culturally based concepts of sickness and health, as well as behavioral norms, allows you to work with your client to set goals that are both personally and culturally acceptable. Finally, attention to cultural factors will help you select the treatment modality most congruent with your client's ethnicity, as well as enabling you to modify traditional treatment approaches as necessary to meet your client's cultural needs.

Sue and Sue (1990) write that certain culture-bound values present sources of conflict and misinterpretation within the context of counseling. Attention to these cultural values will help promote effective communication between counselor and client. We suggest that the following factors, based on the ideas of Sue and Sue (1990), deserve careful consideration as you work to develop a multicultural perspective:

- Individualism versus group orientation
- Verbal and emotional expressiveness versus reservedness

- Valuing insight, self-exploration, or self-knowledge versus valuing "not thinking too much" or "avoiding disturbing thoughts"
- Openness to intimacy versus reluctance for self-disclosure
- Analytic, linear thinking versus intuitive thinking
- Separation of mind and body versus close connection between mind and body
- Expectation for social structure versus tolerance for ambiguity in social situations
- Open patterns of conversation versus communication patterns organized according to social dominance and deference*

## TREATMENT MODALITIES

### Psychodynamic Therapies

Classic psychoanalytic therapy requires a higher level of training for you and a larger investment of time and money for your client than is realistically possible during your internship. However, many important concepts and views introduced by psychoanalytic and psychodynamic theorists underpin other models of therapy as well as our current understanding of human behavior. We feel that the psychodynamic ideas and approaches described in this section apply to your work and may prove useful to you as a neophyte counselor.

Corey (1991a) writes that psychodynamic thinking provides "a focus on the past for clues to present problems" (p. 47). As a counselor intern, you can examine your client's developmental history to gain a more thorough understanding of his or her current level of functioning. Taking the psychosocial history is an important part of your client assessment and is discussed in more detail in Chapter 5.

Psychodynamic concepts of transference and countertransference provide useful avenues for understanding and helping your clients during your internship. In counseling, *transference* is a process whereby the client unconsciously transfers certain aspects of past or present interpersonal relationships onto the current relationship with the counselor. Transference refers to the client's unconscious tendency to experience feelings, attitudes, longings, and fears toward the counselor that were originally felt for other important people in the client's life (Kahn, 1991). For example, a woman who was raised in a family where she was not allowed to express any anger toward her parents may begin to feel and express strong anger toward her counselor during every counseling session. Transference also includes the client's unconscious tendency to attribute attitudes to the counselor that in reality are, or were, held by other important

*Adapted from *Counseling the Culturally Different: Theory and Practice*, 2nd Edition, by D. W. Sue and D. Sue. Copyright © 1990 by John Wiley & Sons, Inc. Reprinted by permission.

people in the client's life. For example, the client whose wife is highly critical of him may begin to perceive the counselor as critical when actually the counselor has no feelings of disapproval for the client.

*Countertransference* refers to the feelings evoked in the counselor by the client, which may be due to the counselor's own issues or to the client's behavior. Countertransference has come to be broadly interpreted to include all the reactions or feelings the counselor experiences toward the client (Kahn, 1991). Countertransference is useful to you as a counselor intern because it may help you understand the impact your client has on other people, which in turn affects the client. For example, if you find yourself feeling constantly annoyed by the client, you may understand a bit more why the client complains that people always avoid him or her. On the other hand, you may be annoyed by the client who reminds you of a problematic issue of your own. Self-knowledge and awareness of your own issues are essential so that you can monitor your countertransferential feelings and be certain that they do not interfere with your effectiveness as a counselor.

Although psychodynamic therapists use transference extensively as a therapeutic means to allow clients to work through their problems, we suggest that simply understanding the concept of transference may allow you to be more helpful to your client during your internship. For example, if your client expresses hurt feelings because you are acting "cold" during a session, when you in fact have warm feelings for the client, you should first examine your behavior to be certain you have not really done or said anything to give this client the feeling that you are acting cold. Next, you should empathize and express to the client how uncomfortable it must be to feel that the counselor is cold. Finally, you should explore the client's feelings by asking about your counseling relationship in the present and also by inquiring whether the client has ever felt that other people have treated him or her coldly (Kahn, 1991).

The psychodynamic term *object relations* refers to "interpersonal relations as they are represented intrapsychically" (Corey, 1991b, p. 112). Freud uses the word *object* to refer to the significant person or thing that is the object of our feelings, wishes, or needs. Corey (1991b) explains that the object relations of early life are replayed throughout the life cycle, even in adult interpersonal relationships, because people unconsciously seek reconnection with their parents and therefore repeat early childhood patterns of interaction. In other words, we all tend to repeat, to some extent, the interpersonal dynamics of our family of origin. You can use this psychodynamic insight to encourage your clients to become aware of repetitive patterns in relationships. Often, a client can be helped to change a potentially destructive behavior by understanding where the behavior is originating. For example, a man who chooses emotionally distant, critical partners over and over again can be helped to understand that he may be recreating his own early, painful relationship with an unavailable, rejecting parent. Together, you and this client can then work on issues such as self-esteem, anger, shame, loss, and improved coping skills so that he does not need to be so stuck in these damaging relationships.

A familiarity with the defense mechanisms described in the psychodynamic literature will serve you well as you work to understand clients. We suggest that you take some time to review these and remember also that your client will use a defense mechanism unconsciously and automatically for protection when he or she is feeling overwhelmed, threatened, or otherwise unsafe (Clark, 1991; Satir, 1987). Your client's use of defense mechanisms is a signal for you to notice. Perhaps the material is too difficult, and you need to slow down or wait before addressing it. On the other hand, perhaps the client is avoiding the reality of painful issues and needs to be confronted or encouraged to explore the material.

Heinz Kohut (1977), a psychodynamic theorist, writes that human beings require that three basic needs be met in order to develop a complete sense of self. These are:

- The need for mirroring, which means being valued and approved of.
- The need for an idealized other, which means having another person to turn to for comfort and safety.
- The need for belonging, which means feeling like other people and being accepted as part of a group.

Kohut (1977) explains that we have these needs throughout our lifetimes and that often our sense of well-being is determined by our ability to find other people, whom he terms *self-objects*, who can meet our needs. Although he writes that these needs are ideally met by parents during early childhood, he suggests that our needs are ongoing and that an incomplete sense of self can be repaired even in adulthood if the needs are met by a caring other, such as a therapist. Kohut's ideas, which form the basis for his "self psychology," have significance for you as a counselor intern in three ways. First, recognizing the underlying unmet needs that are causing your clients pain helps you to be more empathic. Second, you may be more able to easily understand what sort of therapeutic relationship, response, or approach would be most healing for each client in the light of his or her unmet needs. This allows you to focus your interventions more effectively. Third, you can provide hope and a sense of empowerment to your clients by helping them understand that their needs and yearnings are universal, and that they can learn to establish healthy relationships throughout their lives.

## Cognitive-Behavioral Therapies

Cognitive-behavioral therapies apply to many diverse counseling settings, and the underlying concepts, as well as most of the techniques generated by these concepts, are well-suited for your use during your internship. Cognitive-behavioral teachings stress the role of thinking in influencing our sense of well-being and in controlling our mental health. Cognitive-behavioral theorists maintain that emotions and behaviors are directly linked to cognitive processes, so that the way we think about ourselves or the world determines how we feel and how we act (Corey, 1991a, 1991b). In cognitive-behavioral therapy, the clients' dysfunctional thought patterns and their maladaptive interpretations and conclusions about themselves and their environments are examined, tested against contradictory evidence, challenged, and replaced with more adaptive beliefs, leading to therapeutic change (Beck & Weishaar, 1989). As a counselor intern, you will often find these concepts to be useful in empowering your clients because you will be demonstrating to them that their own thoughts are largely responsible for their emotional pain or behavioral problems. Therefore clients can learn to help themselves feel better and make changes by controlling their own thoughts.

Albert Ellis (1989) describes his A-B-C theory, in which he identifies irrational beliefs, rather than actual circumstances or events, as the causative factors in how we feel and act. He writes that *A* represents the activating event, *B* represents our beliefs about the event, and *C* represents the consequences evoked by our beliefs. Typically these consequences are feelings and actions. This theoretical concept may be stated more simply as "It depends on how you look at things" or "You can picture the glass as either half empty or half full."

During your internship, you may be able to use a simple story to illustrate the A-B-C link between thinking and feeling for your clients. For example, we ask our clients to imagine being awakened at 5 A.M. by the sound of someone opening and closing the front door. We ask them how they would feel if they thought an intruder had broken into their home, and most clients answer that they would feel afraid and anxious. Next we ask them how they would feel if they were awakened by the same sound at 5 A.M. but remembered that a family member had planned to get an early start for work. Most clients respond that they might feel calm or relieved at that thought but would not feel afraid or anxious. We discuss with our clients the difference in feelings evoked by different thoughts about the same sound.

As a counselor intern, you can use *cognitive restructuring* with your clients by helping them identify their basic irrational beliefs and automatic negative thoughts. You can help them to think things through in a more rational, positive manner. Often you will be helping them explore alternative attributions for other people's behaviors as well. For example, perhaps a client reports feeling depressed because a friend has not telephoned all week, and therefore the client assumes that the friend no longer likes him or her and feels unattractive and unlovable. As a counselor intern, you may ask your client to try to imagine other explanations for the friend's delay in calling. Could this friend have had a busy week? Been out of town? Tried to call but got a busy signal or missed finding the client at home? Simply forgot to call?

Cognitive-behavioral theorists advocate using questions and a collaborative stance so that you and the client are a team, working together. You may ask your client what the realistic possibility is that the friend actually still likes him or her, even if there was no phone call all week. You can go on to question whether the friend's view of this client actually changes the client's self-worth. After all, the client remains the same, regardless of this particular friend's thoughts. You can then generalize this concept, helping the client challenge his or her assumption that self-worth depends on any other person's evaluation.

When the client begins to understand that the friend may not have called for several possible reasons, that the friend may in fact still like the client despite not calling, that the client still has value and is not any less worthy even if the friend or any other person no longer likes him or her, the client will begin to feel less depressed. His or her maladaptive thoughts led to the depressed feelings; restructuring those thoughts leads to an improved mood. With practice, your client can learn how to think things through even outside your counseling sessions. Cognitive–behavioral therapists call this technique *self-talk*.

Another effective cognitive–behavioral technique is called *catastrophizing*. In practical terms, that means asking your client to verbalize the worst possible scenario and then to explore consequences and coping mechanisms. This is another way of teaching clients to think things through rationally and logically rather than being controlled by their thoughts. For example, one of our clients, a 45-year-old accountant, told us she felt tense during her counseling sessions and exhausted afterward. She related this feeling to trying so hard not to cry during her sessions. Her irrational belief was "I should be able to control my feelings because I'm a grown woman." Closely connected to this thought was the assumption that her counselor would think she was acting like a baby if she cried. After empathizing with this client's feelings and acknowledging the hard work she had been doing to keep from crying, we assured her that we believed crying during her counseling session was acceptable and probably helpful, rather than a sign of being out of control or immature. We asked her to imagine what would happen if she did allow herself to cry (verbalizing the worst scenario: her crying during the session). "I probably couldn't stop, once I got started," she said (exploring consequences). However, as she thought things through, she decided that even if she cried for a long time, she would eventually be able to stop (exploring coping mechanisms), and that she might feel relieved and be able to make better use of her sessions if she did not have to struggle to maintain her composure (thinking things through realistically).

Corey (1991a) points out possible pitfalls of cognitive–behavioral therapies: their focus on thinking may neglect exploration of emotional issues, and they may lead to too much intellectualization by both counselors and clients. He advises that this type of therapy is not suitable for clients of limited intellectual ability. In addition, he suggests that counselors may too often tend to impose their own values and views on clients. Corey warns about "the therapist who beats down clients with persuasion" (p. 143) when using cognitive–behavioral techniques. We would like to add that clients can easily misinterpret your efforts to challenge their irrational beliefs and see you in a punitive, disapproving, or parental role. As a counselor intern, you need to take great care to be empathic, to maintain your therapeutic relationship, and to emphasize a collaborative approach when you use cognitive–behavioral treatment modalities.

## Humanistic/Existential Therapies

Humanistic/existential schools emphasize a phenomenological, experiential approach to counseling. These therapies, which draw on spiritual beliefs and existential philosophies as well as psychological principles, stress the quality of the therapeutic relationship itself, rather than a set of therapeutic techniques, as an avenue toward client self-awareness and change (Corey, 1991a). The counselor's abilities to grasp the client's subjective internal experience and to relate to the client in an authentic manner are considered to be of paramount importance (Lewis, Hayes, & Lewis, 1986).

Carl Rogers suggests that all human beings have an actualizing tendency toward growth, health, and the fulfillment of potential, and that this tendency can be nurtured by the provision of a relationship characterized by genuineness, unconditional positive regard, and empathy (Raskin & Rogers, 1989). Rogers's teachings have two very significant implications for you during your internship. First, Rogers's belief in an inherent, universal human potential for self-actualization means that, as an intern, you do not need to be too concerned about diagnostic categories or labels in order to decide how to build a relationship with clients. You will be able to approach all clients in the same way. Second, you will be assured that even though you have relatively little clinical experience or counseling practice, you will be able to help your clients by interacting with them in an authentic, accepting, and empathic manner.

*Genuineness,* Rogers's first characteristic of the optimal therapeutic relationship, means being in touch with your own inner experiences and feelings and presenting yourself as a real person, rather than retreating behind a professional facade. Being genuine, or congruent, means neither that you should reveal all aspects of your personality to your client, nor that you should express every thought and feeling that you experience. Rogers (cited in Baldwin, 1987) explains:

> The self I use in therapy does not include all my personal characteristics. Many people are not aware that I am a tease and that I can be very tenacious and tough, almost obstinate.... I guess that all of us have many different facets, which come into play in different situations. I am just as real when I am being understanding and accepting, as when I am being tough. To me being congruent means that I am aware of and willing to represent the feelings I have at the moment. It is being real and authentic in the moment. (p. 51)

Shea (1988) further clarifies the term *genuineness* by noting that being genuine, for the clinician, means being responsive, being appropriately spontaneous, and being consistent in relating to the client.

The second aspect of Rogers's therapeutic relationship is *unconditional positive regard.* Rogers writes that this concept refers to the counselor's caring and nonjudgmental acceptance of all the client's feelings and thoughts; the counselor must respect and prize the client (Raskin & Rogers, 1989). As an intern, you will at times counsel clients whose attitudes, values, feelings, and behaviors conflict with your own beliefs and ideals. Your goal and responsibility in these situations is maintaining your unconditional positive regard for the client as a unique and valuable human being.

*Empathy* is the third relational quality that Rogers finds to be essential in counseling. Empathy may be conceptualized as the counselor's ability to be emotionally attuned to the client, or the counselor's ability to recognize, understand, and share the client's internal world. Many authors (Gilbert, 1992; Kahn, 1991; Raskin & Rogers, 1989; Shea, 1988) point out that empathic healing is actually an interpersonal, interactive process. The counselor must not only be empathic but also be able to convey that empathic caring and interest so that the client is able to perceive it clearly. Gilbert (1992) describes a model for this empathic process: "I understand. I show you I understand, and you understand that I have understood" (p. 11).

Kahn (1991) summarizes the therapeutic value of an empathic connection between counselor and client by writing, "When I really get it that my therapist is trying to see my world the way I see it, I feel encouraged. . . . If my therapist thinks it's worth the time and effort to try to understand my experience, I must be worth the time and effort" (p. 43). As a counselor intern, therefore, one of your most important tasks in helping your clients will be working hard to communicate your empathy, interest, warmth, and caring and to make certain that your clients are aware that you have these feelings for them.

Gestalt therapy is similar to Rogers's client-centered approach in its emphasis on client self-awareness and growth facilitated by participation in an authentic counseling relationship. Gestalt therapy is experiential and stresses client exploration and integration of thoughts, emotions, and behaviors through awareness of the here-and-now (Corey, 1991a, 1991b). The client and the counselor work to compare their individual phenomenological perspectives, exploring their differing viewpoints with the goal of increasing client awareness, insight, acceptance, and responsibility (Corey, 1991a; Yontef & Simkin, 1989).

Gestalt therapy places more emphasis on specific counseling techniques, termed *experiments* or *exercises*, than do the other humanistic/existential approaches. One Gestalt technique that we have found useful to counselor interns involves helping your client to identify and focus on the feeling, for example, anger, that is beneath his or her nonverbal behavior, such as clenched fists, and then asking your client to *stay with the feeling*. As the client experiences the feeling fully, he or she often explores important issues and gains self-awareness.

Another Gestalt technique you may find useful during your internship is the *empty chair*. Here, clients imagine someone, such as a parent or spouse, to be sitting in a nearby chair. Clients then say all the things they wish they could really tell that person. This directed verbalization often is cathartic for the client and leads to new understandings for both client and counselor.

Many Gestalt techniques will be useful during your internship because they will help you attend closely to the significance of your clients' nonverbal behaviors as well as their use of language in relation to their thoughts and feelings. As a counselor intern, you will be able to understand your clients more fully and grasp the depths of their feelings and experience by attending carefully to their nonverbal signals and cues. For example, noticing your clients' muscle tension, body posture, and hand and facial movements gives you a great deal of information. In addition, you will be able to understand and empathize with your clients' perceptions of self and world by listening closely to the words they select to express themselves. For example, the client who says, "He makes me so depressed," feels less powerful and assumes less responsibility than the client who says, "I feel depressed when he does that."

Corey (1991a, 1991b) cautions that the humanistic/existential therapies focus on the present moment to such a great extent that the influence of the client's developmental history and the power of the past may be too often discounted. In addition, he suggests that the emphasis on emotional facets results in less attention to the importance of cognitive or intellectual factors in understanding clients.

## ADJUNCT TREATMENT MODALITIES

### Group Therapy

You will most likely be asked to facilitate or co-facilitate at least one group as part of your internship training. Group counseling is used in a wide range of settings, with diverse populations, and for varied purposes. Groups generate powerful interpersonal forces that may be used for education, support, and prevention or remediation of psychological issues. Corey (2000) identifies the following general goals of group counseling: learning to trust self and others, recognizing commonalities of problems and needs, achieving greater self-awareness, increasing self-confidence and self-acceptance, finding alternative ways to manage problems and resolve conflicts, and becoming more aware of one's responsibility toward self and others. Yalom (1985) writes that the therapeutic factors of group counseling may include altruism, universality, ventilation of feelings, instillation of hope, and improved interpersonal skills.

When you facilitate groups during your internship, you can focus your interventions toward accomplishing these goals, encouraging these therapeutic factors to come into play, and helping individual members to use the group's healing forces. In concrete terms, that means:

- Ensuring a safe place in the group by protecting members physically and psychologically.
- Providing appropriate structure and clear rules so that a therapeutic framework is maintained.
- Encouraging direct communication and interaction among group members.
- Modeling active listening skills.
- Providing positive feedback when group members take risks, work hard, show increasing self-awareness, offer support for others, and so on.
- Being aware of the developmental stages of group process.
- Asking questions such as "How many others have felt that way?" or "How many others have had a similar experience?"
- Helping members become aware of and express their current feelings concerning group interaction.

## Bibliotherapy

Bibliotherapy involves using books or other reading material to help your clients. A book may be of the popular self-help variety, a booklet prepared especially for clients with a particular problem, or even a professional resource that presents relevant information without using too much jargon. You may suggest that your client read outside your sessions, and then you may choose to discuss the books with your client during the session or read an appropriate passage together. Some groups are structured so that members take turns reading aloud, followed by a discussion of personal reactions to each section. In each situation, we suggest that you first read the books yourself to be sure that the material is accurate and appropriate for your client's unique concerns and current stage in the counseling process. In addition, you can use bibliotherapy most effectively by helping your clients explore their own thoughts and feelings about the reading material. Often bibliotherapy not only helps clients by providing factual information but also offers some of the same therapeutic factors as group therapy, including universality, guidance, identification, self-understanding, and instillation of hope.

## Art Therapy

Neither you nor your client needs any special artistic talents to use art during your counseling sessions. Art therapy is an alternative means of nonverbal communication that you can use to help clients during your internship. We tell clients that "art helps us say things that sometimes just can't be put into words." In using art with clients, keep in mind that any artwork produced during your counseling session is likely to be a very important, intensely personal representation of the client's sense of self or inner world because feelings and issues are being stirred up by the treatment situation. Be sure to demonstrate your interest, caring, and warm acceptance of the artwork. Also, take time to explore the client's feelings and thoughts by talking about the artwork.

We suggest using simple colored markers and a 9- by 12-inch pad of white paper during your internship. These are easy to carry with you, do not require any special space or preparation, and are nonthreatening and easy for most clients to use. You may find it helpful to explain as you offer the art materials that "this is about feelings, not really about the art being good or bad." Many clients are hesitant to begin and may need some encouragement or direction. If this is the case, you can offer one or two specific suggestions about what to draw. Possible topics include an event from the past, a dream, the client's family of origin, how the client is feeling right now, a wish, a favorite activity, a fear, an important person in the client's life, or a self-portrait.

When a client draws an upsetting or frightening picture, you can help him or her experience a sense of power and control by suggesting ways to change the drawing to make it safer or less scary. For example, one young client became anxious and afraid after drawing a picture of a large dog that had attacked and bitten her. She felt safer when she drew a strong cage around the dog and added a lock on the cage door. Another client who had been mistreated by someone experienced a sense of relief when she chose to draw a picture of him, rip it up, and throw the pieces of paper in the trash.

Art is a powerful therapeutic tool, and we have found it to be helpful in the following ways:

- Art provides a cathartic experience; strong emotions are released as art is produced.

- Art increases the connection between counselor and client as the artwork is shared and discussed.

- Art helps increase client self-esteem. The counselor values and accepts the client's artwork and, in so doing, demonstrates that the client is also worthy of being valued and accepted.

- Art is empowering for the client, who directly manipulates his or her environment while having control of the creative process.

## CONCLUDING REMARKS

Michael Kahn (1991) writes that as counselors, "We are not doing therapy the way a surgeon does surgery; we are the therapy" (p. 44). During your internship, we hope that you will value yourself as your most powerful means of helping your client. Studies (Rubin & Niemeier, 1993) have demonstrated that successful outcome in psychotherapy is positively correlated with two factors: (1) the strength of the relationship between counselor and client and (2) the client's perception of the counselor's warmth and empathy. As a beginning counselor, your particular theoretical orientation and the specific treatment modalities you choose are not as important as your caring, your sincere efforts to understand your client, and your ability to convey to your client your empathy, respect, and desire to be helpful.

In this chapter we have included an overview of some of the concepts and philosophies underlying the three major theoretical schools of counseling and psychotherapy, as well as a discussion of the various treatment modality aspects we feel are most relevant to you as a counselor intern. We have limited our choices not only with regard to the constraints of the time and space of one chapter of a single book, but also with respect to your level of training as an intern. As you gain clinical experience, become more comfortable interacting with clients, encounter more difficult problems, refine your personal counseling style, and function with more autonomy in your counseling career, we encourage you to enrich your repertoire of treatment modalities and counseling methods to help your clients reach their highest levels of functioning.

# REFERENCES

BALDWIN, M. (1987). Interview with Carl Rogers on the use of self in therapy. In M. Baldwin & V. Satir (Eds.), *The use of self in therapy* (pp. 45–52). New York: Haworth.

BECK, A., & WEISHAAR, M. (1989). Cognitive therapy. In R. Corsini & D. Wedding (Eds.), *Current psychotherapies* (4th ed., pp. 285–322). Itasca, IL: Peacock.

CARKHUFF, R. (1984). *Helping and human relations* (Vol. 2). New York: Holt, Rinehart, and Winston.

CLARK, A. (1991). The identification and modification of defense mechanisms in counseling. *Journal of Counseling and Development, 69,* 231–235.

COREY, G. (199la). *Manual for theory and practice of counseling and psychotherapy.* Pacific Grove, CA: Brooks/Cole.

COREY, G. (199lb). *Theory and practice of counseling and psychotherapy* (4th ed.). Pacific Grove, CA: Brooks/Cole.

COREY, G. (2000). *Theory and practice of group counseling* (5th ed.). Pacific Grove, CA: Brooks/Cole•Wadsworth.

DACEY, J., & TRAVERS, J. (1991). *Human development across the lifespan.* Dubuque, IA: Brown.

DOYLE, R. (1998). *Essential skills and strategies in the helping process* (2nd ed.). Pacific Grove, CA: Brooks/Cole.

ELLIS, A. (1989). Rational emotive therapy. In R. Corsini & D. Wedding (Eds.), *Current psychotherapies* (4th ed., pp. 197–240). Itasca, IL: Peacock.

GABBARD, G. (1990). *Psychodynamic psychiatry in clinical practice.* Washington, DC: American Psychiatric Press.

GIBBS, J., & HUANG, L. (1989). A conceptual framework for assessing and treating minority youth. In J. Gibbs & L. Huang (Eds.), *Children of color* (pp. 1–29). San Francisco: Jossey-Bass.

GILBERT, P. (1992). *Counseling for depression.* London: Sage.

GUTHEIL, T., & GABBARD, G. (1993). The concept of boundaries in clinical practice: Theoretical and risk-management dimensions. *American Journal of Psychiatry, 150*(2), 188–196.

KAHN, M. (1991*). Between therapist and client.* New York: Freeman.

KOHUT, H. (1977). *The restoration of the self.* New York: International Universities Press.

KURPIUS, D. (1986). The helping relationship in counseling and consultation. In M. Lewis, R. Hayes, & J. Lewis (Eds.), *The counseling profession* (pp. 96–129). Itasca, IL: Peacock.

LEE, C. (1991). Cultural dynamics: Their impact in multicultural counseling. In C. Lee & B. Richardson (Eds.), *Multicultural issues in counseling* (pp. 11–22). Alexandria, VA: American Counseling Association.

LEWIS, M., HAYES, R., & LEWIS, J. (1986). *The counseling profession.* Itasca, IL: Peacock.

LOPEZ, S., GROVER, K., HOLLAND, D., JOHNSON, M., KAIN, C., KANEL, C., MELLINS, A., & RHYNE, M. (1989). Development of cultural sensitivity in psychotherapists. *Professional Psychology: Research and Practice, 20*(6), 369–376.

RASKIN, N., & ROGERS, C. (1989). Person-centered therapy. In R. Corsini & D. Wedding (Eds.), *Current psychotherapies* (4th ed., pp. 155–196). Itasca, IL: Peacock.

ROGERS, C. (1951). *Client-centered therapy.* Boston: Houghton Mifflin.

ROGERS, C. (1957). The necessary and sufficient conditions of therapeutic personality change. *Journal of Counseling Psychology, 21*(2), 95–103.

ROGERS, C. (1961). *On becoming a person.* Boston: Houghton Mifflin.

ROOT, M. (1985). Guidelines for facilitating therapy with Asian American clients. *Psychotherapy, 22,* 349–356.

RUBIN, S., & NIEMEIER, D. (1993). Non-verbal affective communication as a factor of psychotherapy. *Psychotherapy, 29*(4), 596–602.

SATIR, V. (1987). The therapist story. In M. Baldwin & V. Satir (Eds.), *The use of self in therapy* (pp. 17–26). New York: Haworth.

SHEA, S. (1988). *Psychiatric interviewing: The art of understanding.* Philadelphia: Saunders.

SUE, D., & SUE, D. (1990). *Counseling the culturally different.* New York: Wiley.

VELLEMAN, R. (1992). *Counseling for alcohol problems.* London: Sage.

VONTRESS, C. (1986). Social and cultural foundations. In M. Lewis, R. Hayes, & J. Lewis (Eds.), *The counseling profession* (pp. 215–250). Itasca, IL: Peacock.

YALOM, I. (1985). *Theory and practice of group psychotherapy.* New York: Basic Books.

YONTEF, G., & SIMKIN, J. (1989). Gestalt therapy. In R. Corsini & D. Wedding (Eds.), *Current psychotherapies* (pp. 328–362). Itasca, IL: Peacock.

# BIBLIOGRAPHY

BASCH, M. (1980). *Doing psychotherapy.* New York: Basic Books.

BASCH, M. (1988). *Understanding psychotherapy.* New York: Basic Books.

BECK, A., & EMERY, G. (1985). *Anxiety disorders and phobias: A cognitive perspective.* New York: Basic Books.

CHIAFERI, R., & GRIFFIN, M. (1997). *Developing fieldwork skills.* Pacific Grove, CA: Brooks/Cole.

DRYDEN, W., & HILL, L. (1993). *Innovations in rational emotive therapy.* London: Sage.

KOTTLER, J. (1990). *On being a therapist.* San Francisco: Jossey-Bass.

KOTTLER, J. & BROWN, R. (1996). *Introduction to therapeutic counseling* (3rd ed.). Pacific Grove, CA: Brooks/Cole.

MARTIN, D., & MOORE, A. (1995). *First steps in the art of intervention.* Pacific Grove, CA: Brooks/Cole.

NICKERSON, E. (1983). Art as a play therapeutic medium. In C. Schaefer & K. O'Conner (Eds.), *Handbook of play therapy* (pp. 234–250). New York: Wiley.

OSTER, G., & GOULD, P. (1987). *Using drawings in assessment and therapy.* New York: Brunner/Mazel .

SEINFELD, J. (1993). *Interpreting and holding: The paternal and maternal functions of the psychotherapist.* Northvale, NJ: Jason Aronson.

SELIGMAN, L. (1990). *Selecting effective treatments.* Alexandria, VA: American Counseling Association.

STORR, A. (1990). *The art of psychotherapy* (2nd ed.). New York: Routledge.

URSANO, R., SONNENBERG, S., & LAZAR, S. (1991). *A concise guide to psychodynamic psychotherapy.* Washington, DC: American Psychiatric Press.

VRIEND, J. (1985). *Counseling powers and passions.* Alexandria, VA: American Counseling Association.

WACHTEL, R. (1993). *Therapeutic communication: Principles and effective practice.* New York: Guilford.

WRIGHT, J., THASE, M., BECK, A., & LUDGATE, T. (Eds.). (1993). *Cognitive therapy with inpatients.* New York: Guilford.

# 5

# The Clinical Interview

Many interns begin their experience by performing a clinical interview as part of the intake process. Note our intentional use of the word *process*. We view the intake as an important component of the therapy itself, rather than as a distinct piece of "busy work" apart from the dynamics of counseling. In other words, the clinical interview starts establishing the therapeutic relationship between the counselor and client, which conveys a sense of client worth and worthiness for treatment. It has been the authors' experience that in many agencies and institutions clinicians view the intake as tedious, somewhat routine, and requiring minimal skills. As a result, interns or novice clinicians often perform the intake. Generally, the intake is the first contact many clients have with the mental health system and with therapy and counseling itself; therefore, it sets the tone for treatment, for sharing agency and counselor philosophy and values, and for treatment expectations.

The many types and methods of intake depend mostly on agency policy and style. Some are rather informal, calling for minimal information. Others are formalized procedures of data gathering involving the completion of several intricate forms, some by the client, others by the counselor. All share in the goal of helping clients successfully empower themselves to address presenting problems. Refer to Appendixes H and I for sample intake and treatment plan forms.

We view the intake as a systematic process of information gathering and sharing that usually includes the following areas: (1) client description, (2) problem description, (3) psychosocial history, (4) mental status examination, (5) diagnostic impression, and (6) treatment recommendations (Faiver, 1988). Further, as Benjamin (1987) notes, the counselor-in-training should view the counseling relationship as special, thereby ensuring full concentration of energy on the client and his or her situation and unique personhood.

The client should be encouraged to ask questions and share expectations regarding treatment. Also, we suggest that you consider discussing what counseling and therapy are and are not. For example, counseling and therapy are a mutual process of personal exploration, growth, hard work, and anticipated problem resolution, not exercises in magical advice giving and unilateral direction by the counselor. Further, counselors should allow sufficient time to conduct the interview based upon the presenting problem, client's history, level of present functioning, and the client's willingness to engage in this process. Thus, conducting a comprehensive clinical interview as a foundation for successive therapeutic interventions may take more than one session.

At this juncture, let's describe in detail each of the areas of the formal intake process.

## CLIENT DESCRIPTION

Acquiring basic background information on your client sets the initial phase of the clinical interview. With these data, you establish a general framework by which you initiate a relationship with the client. You will soon realize this knowledge fits into your understanding of key developmental influences of the client's life. We suggest the following information be gathered:

- Name
- Address
- Telephone number
- Emergency contact person with telephone number
- Guardian and/or responsible party involvement for minors
- Birthdate
- Gender
- Race
- Nationality
- Marital status
- Children
- Religion
- Level of education (including regular or special)
- Financial data (including Medicaid, Medicare, self-pay, or insurance information)
- Socioeconomic information

- Cultural influences
- Present job/position
- Referral source

Once this information is collected, you may already have formulated some beliefs about the individual and the direction to take in counseling.

## PROBLEM DESCRIPTION

You must encourage your client to define his or her problem or concern at that particular time and be aware that explanation of this problem area is from his or her sole vantage point. As the interview progresses, you may discover incongruities or discrepancies that may need sorting out at some point. Delineating the scope of the problem and its effects on the client's and significant others' present levels of functioning is most relevant and a key to the flow of the counseling. Inquire about the client's perception of the etiology of the problem, such as when the problem started and any influencing interpersonal and environmental factors. Even though clients present a specific uniqueness to their problems and are unique individuals in and of themselves, you need to convey some assurance that you have dealt with similar concerns, maintaining both your clinical objectivity and respect for the clients' individuality. Clients usually expect that you will be able to state your understanding of the problem area (Sullivan, 1970). This statement reassures clients and enhances the development of the therapeutic alliance and subsequent therapeutic journey; it also can facilitate the direction of treatment. If you are not sure exactly what the defined problem is, ask and get clarification so as not to impede the direction of the interview. Frequently, the presenting problem may mask a significant underlying issue, which may not surface until trust is established. Further, you can explore the history of similar problems and how they were resolved.

## PSYCHOSOCIAL HISTORY

Now that the problem area has been designated, explore the client's background by taking the psychosocial history. Sullivan (1970) discusses the importance of gathering a "rough social sketch" (p. 72) of the client rather than an extensive and detailed life history. Your psychosocial history should address the developmental milestones to the present. MacKinnon and Michels (1971) believe that the psychosocial history has more diagnostic and treatment value than physical examinations or laboratory studies. Unlike Sullivan (1970), they choose to focus on an extensive developmental history from infancy to adulthood. According to these authors, being knowledgeable about the developmental milestones of each stage is relevant. In addition, they assert that the history conveys the strengths and weaknesses of the client in relation to the various interpersonal and environmental factors of his or her life.

We follow this format. As with any outline, it should be adjusted to meet your needs.

- Developmental history—emphasizing milestones and anomalies of client development
- Description of client's family—mother, father, brothers, sisters, and any significant others; description of client's relationships with these individuals
- Description of early home life (happy, abusive, or the like)
- Educational history
- Occupational history
- Social activities—dating, sexual history, relationship history, and problems
- Marital history—including separations, divorces
- Health history—physical, mental, accidents, illnesses, surgeries, hospitalizations, psychiatric medication (past and present), and other medications (past and present)
- Substance use, abuse, and dependence
- Legal history including probation and parole
- Present living arrangements
- Significant life events
- Lowest points of the client's life
- Abuse issues—physical, emotional, and sexual
- Psychiatric history, inpatient and outpatient
- Individuals who had greatly influenced the client's life
- Who influences the client presently
- Past and present support systems (such as parents, friends, or spouse)
- Ethnic/cultural influences
- Religious history and present belief system
- Community involvement, interests, and leisure activities
- Military history, including type of discharge
- Significant environmental influences, such as school, political, religious
- Age-specific issues in minors
- Anything else we should know but failed to ask

We suggest that you explore in depth any specific area of history that has the most relevance to present problem areas and functioning.

# MENTAL STATUS EXAMINATION

The mental status examination is your stethoscope for understanding the client's behavioral, cognitive, and affective domains. In fact, Siassi (1984) notes, "in the psychiatric evaluation the mental status examination is considered to be analogous to the physical examination in general medicine" (p. 267). This evaluation is a key factor in determining a diagnostic impression.

You may want to intersperse the various components of the mental status evaluation throughout your interview. This approach tends to be more comfortable for clients because it avoids the drudgery of a long list of questions.

Faiver (1988) focuses his mental status evaluation on assessing (1) behavior, (2) affect, and (3) cognition. We advise the intern to place equal emphasis on all three areas but to feel comfortable, when necessary, delving into any of them for further clarification. In addition, order of assessment should be based on personal preference and style. We focus briefly on each area.

## Behavior

Evaluate how the client's behavior affects present functioning. Note overall appearance, including specific physical characteristics, unusual clothing, mannerisms, gestures, tics, motor activities, posture, level of self-care, level of activity, sleeping and eating disturbances, fatigue symptoms, physical or verbal aggression, and the client's reaction to you as therapist. What does the client's body language convey to you? Does his or her posture denote an open or closed stance? Are there any psychosomatic issues? Observe whether the client can attain and maintain eye contact. Note any changes of facial expression or any other noteworthy behaviors.

## Affect

Next concern yourself with the client's feelings. Evaluate the client's overall pervasive mood. Is the client's overall feeling level dysphoric (depressed), irritable, expansive, elevated, or euphoric? Note the degree of affect. Are the client's feeling levels mild, moderate, or severe? Can he/she identify and label his or her feelings? Does the client experience mood fluctuation? At times counselors must clarify ambiguous feelings to ensure common understanding.

## Cognition

Perhaps this section of the mental status evaluation is the most comprehensive because it focuses on the greatest number of variables. Note, however, that all areas of assessment in this section deal with the client's former or current thought processes. Here the counselor observes the client's alertness and responsiveness. Orientation in the three spheres (person, place, and time) is assessed. In other words, does the client know who he or she is? where he or she is? and what the date and time are? Is he or she thinking abstractly or concretely? For example, is the adult client's thinking regressed and compromised?

Assess any delusions (false beliefs that may or may not appear to be realistic) or hallucinations (false perceptions that may be auditory, visual, tactile, or olfactory). Are both recent and remote memory intact? Evaluate attention span. Get an impression of the client's level of intelligence. The client's reality testing can be understood via his or her belief system, the level of insight into his or her problem areas, and whether the client is able to exhibit good judgment. Lastly, question the client on any ideation (thoughts) or plans (intended behaviors) to harm self or others, and note any and all responses verbatim.

As you complete the mental status evaluation, reviewing the three areas of assessment will help solidify your diagnostic impression and treatment recommendations.

## DIAGNOSTIC IMPRESSION

When you arrive at this point in the clinical interview, you are cognizant of the presenting problem, possess a somewhat detailed psychosocial history, and have a good idea of the client's current mental status. The information gathered in the intake interview provides supportive documentation for formulating a diagnostic impression.

Counselors have mixed feelings about diagnosing. After all, is it not labeling? And labeling theory (Becker, 1963) indicates that persons often live up—or down—to a label. Also, one never knows in whose computer banks a diagnosis ends up. Yet, with counseling included in the medical model, we often must diagnose in order to get paid. All insurers and the government require a diagnosis for reimbursement for services. Consider the great ethical and moral responsibility we have!

Licensure and certification laws in most states permit only independent practitioners (psychiatrists, psychologists, clinical counselors, and clinical social workers) to diagnose mental and emotional disorders without supervision. Your responsibility is to develop a diagnostic impression based on your extensive interview. Do not minimize the importance of your diagnostic impression. Making a tentative diagnosis (an impression, not carved in stone) improves with practice, further training, and experience. Finally, diagnostic impressions are fluid based on the client's potential change in level of functioning.

### The *Diagnostic and Statistical Manual of Mental Disorders*

In the mental health field, the diagnostic reference is the *Diagnostic and Statistical Manual of Mental Disorders*, fourth edition, published by the American Psychiatric Association (1994). We suggest that you view this nosology (classification system) as a "cookbook." Each recipe has certain ingredients, which, in certain combinations, produce a certain product—a diagnosis. Each product has a number attached to it for ease of reporting. Diagnoses are clustered by groupings that have some similarities in symptoms. Severity is determined by several factors, including intensity, duration, and type of symptoms.

Be thorough, responsible, and accurate in your diagnosis. Choose labels carefully because, as previously stated, labels are not always used to the client's advantage. Always consult with your site supervisor and other independent practitioner colleagues to verify your impressions.

You may find that your impressions change as you learn more about your client in subsequent sessions and as additional problems are discussed. It is your responsibility to note any modifications to the diagnosis in the client record.

The task of establishing a diagnostic impression must be taken seriously. No two individuals possess the same interpersonal, environmental, or genetic factors. For these reasons your ultimate decision should be individualized even though similarities may be present. We owe it to our clients to assess them in the present moment, with no preconceived notions that may fit a person to a diagnosis. This affords greater objectivity and fairness.

At this point, let's take time to briefly describe the *Diagnostic and Statistical Manual* (DSM-IV) in more detail. As with previous editions, the fourth edition of the manual is compatible with the *International Classification of Diseases Manual* (ICD-10), published by the World Health Organization, which has as its goal the maintenance of a system of coding and terminology for all medical disorders, including those of a psychiatric nature (American Psychiatric Association, 1994). The DSM-IV is the fourth in a series of nosologies published by the American Psychiatric Association (and is actually the fifth revision). Its biopsychosocial approach attempts to consider holistically and atheoretically all possible clinical syndromes and developmental and medical variables of patients and strives to facilitate our diagnostic and treatment planning (American Psychiatric Association, 1994). This is a tall order. And, certainly, the DSM-IV operates from a medical model, which can be limiting to those of us in the mental health field, considering that there are other equally compelling models such as learning and developmental approaches to mental dysfunctions.

Five places or categories, called *axes*, are possible to address in the DSM-IV. They include the following:

Axis I      Clinical Syndromes

            Other Conditions That May Be a Focus of Clinical Attention

Axis II     Personality Disorders

            Mental Retardation

Axis III    General Medical Conditions

Axis IV     Psychosocial and Environmental Problems

Axis V      Global Assessment of Functioning

These are fully described in the DSM-IV as well as other texts. It is not our intent in this book to present a course on the DSM-IV; rather, we wish merely to introduce you to the labeling system of the mental health field. We suggest that you review the books listed in the bibliography and reference sections of this chapter. The DSM-IV may well become a constant companion and a key reference for many of you.

## TREATMENT RECOMMENDATIONS

Treatment recommendations ideally are made in concert with the client. When we include the client in decision making, we begin the teaching aspect of therapy: that counseling is a mutual process in which clients have valued input and equal responsibility regarding their treatment outcomes. Much depends on diagnosis, level of client acuity, client age, issues of convenience, and the expertise, skills, and theoretical orientation of the therapist. Recommendations should be based on what services would facilitate client ability to cope effectively and efficiently with current and future demands and responsibilities. Again, because no two individuals are exactly alike, the decision regarding treatment recommendations is as individualized as the diagnosis.

The following is a treatment recommendation checklist to assist in choice of treatment recommendations:

1. Are the client's problems acute (intense) or chronic (constant over time)?
2. What limitations may impede services (such as finances, transportation, child care, employment, general health, time constraints, or the like)?
3. Does the seriousness of the client's situation demonstrate the need for a structured inpatient environment?
4. Can the client's concerns be handled on an outpatient basis?
5. Are there requirements from the legal system (parole board or probation department)?
6. Are we providing the least restrictive treatment environment that allows the client to address his or her problems?
7. Would the client do better in individual or group counseling?
8. Does the specific diagnosis indicate a greater success rate in a specific treatment modality?
9. Will the client likely need short- or long-term services?
10. Do you need to collaborate with other professionals (such as a psychiatrist, psychologist, or clergy)?
11. Will any restrictions on your time and schedule influence your treatment?
12. Are there presenting problems outside your scope of practice, necessitating transferring the client to another professional within your agency or to another specialized agency? What is the most timely and effective process?

13. Will managed care or an insurance company direct the type and frequency of services?

14. Is your agency equipped with the trained personnel and facilities to address the client's specific problem areas (such as a play therapy room or group therapy room)?

15. Are you able to provide any specific services the client has requested (such as bibliography or hypnosis)?

16. What are the client's strengths and barriers in achieving therapeutic goals?

17. Must any special needs (such as sensory impairments or the need for an interpreter) be addressed?

We attempt to help clients achieve their optimal level of functioning. At times, deciding on treatment recommendations may involve some creativity on our part. Even though we aspire to meet their needs, this may not always be possible. Do the best you can with what resources you have available, always operating ethically. Be candid with clients as to what you can and cannot do. For example, if you discover that neither you nor your agency can address the client's presenting problem, you have the obligation to help that client find the necessary resources (ACA Code of Ethics, section A.11.b; see Appendix E). As always, consult with your site supervisor.

Appendix I provides a sample treatment plan form that follows a management by objective format, is as behaviorally specific as possible, and includes outcome objectives and results. We have found that this sample treatment plan form falls nicely within the parameters of what managed care providers look for in treatment planning. Ideally, we formulate the treatment plan in conjunction with the client; thus the treatment plan becomes a social contract between the counselor and client. Note that we provide signature spaces for both client and counselor (and supervisor). In the case of minors, we ask the minor to sign in addition to the parent. How often are children and adolescents offered the opportunity to sign something? Signatures are one way to enlist investment in therapy. We urge you to discuss your treatment plan with your supervisor.

Upon completion of the clinical interview, clients should feel hopeful and confident in your ability to help them achieve their goals in this, the first step in the treatment process.

## CONCLUDING REMARKS

The intake is a dynamic process involving both counselor and client. Components of the process include client and problem description, psychosocial history, mental status evaluation, diagnostic impression, and treatment recommendations. The ideal is to include the client in every step as a unique and distinct partner in the therapeutic journey.

# REFERENCES

AMERICAN PSYCHIATRIC ASSO-CIATION. (1994). *Diagnostic and statistical manual of mental disorders* (4th ed.). Washington, DC.

BECKER, H. (1963). *Outsiders: Studies in the sociology of deviance*. New York: Free Press.

BENJAMIN, A. (1987). *The helping interview with case illustrations*. Boston: Houghton Mifflin.

FAIVER, C. (1988). An initial client contact form. In P.A. Keller & S. R. Heyman (Eds.), *Innovations in clinical practice: A source book* (Vol. 7, pp. 285–288). Sarasota, FL: Professional Resource Exchange.

MacKINNON, R.A., & MICHELS, R. (1971). *The psychiatric interview in clinical practice*. Philadelphia: Saunders.

SIASSI, I. (1984). Psychiatric interview and mental status examination. In G. Goldstein & M. Hersen (Eds.). *Handbook of psychological assessment*. New York: Pergamon.

SULLIVAN, H. S. (1970). *The psychiatric interview*. New York: Norton.

# BIBLIOGRAPHY

ADLER, A. (1964). *Problems of neurosis*. New York: Harper & Row.

BRAMMER, L.M. (1993). *The helping relationship. Process and skills*. Boston: Allyn & Bacon.

EGAN, G. (1994). *The skilled helper. A problem-management approach to helping*. (5th ed.). Pacific Grove: Brooks/Cole.

EISENGART, S., EISENGART, S., FAIVER, C., & EISENGART, J. (1996). Respecting physical and psychosocial boundaries of the hospitalized patient: Some practical tips on patient management. *Rural Community Mental Health, 23*(2), 5–7.

EISENGART, S., & FAIVER, C. (1996). Intuition in mental health counseling. *Journal of Mental Health Counseling, 18*(1), 41–52.

ELLIS, A. (1989). Rational-emotive therapy. In R. J. Corsini & E. Wedding (Eds.), *Current psychotherapies* (4th ed., pp. 197–238). Itasca, IL: Peacock.

FAIVER, C. (1992). Intake as process. *CACD Journal, 12*, 83–85.

FAIVER, C., & O'BRIEN, E. (1993). Assessment of religious beliefs form. *Counseling and Values, 37*(3), 176–178.

FAIVER, C., O'BRIEN, E., & McNALLY, C. (1998). *Characteristics of the friendly clergy. Counseling and Values, 42*(3), 217–221.

HUTCHINS, D.E., & COLE, C.G. (1986). *Helping relationships and strategies*. Pacific Grove, CA: Brooks/Cole.

PIAGET, J. (1970). The definition of stages of development. In J. Tanner & B. Inhelder (Eds.), *Discussions on child development* (Vol. 4). New York: International Universities Press.

ROESKE, N. (1972). *Examination of the personality*. Philadelphia: Lea & Febiger.

SOMMERS-FLANAGAN, J., & SOMMERS-FLANAGAN, R. (1993). Foundations of therapeutic interviewing. Boston: Allyn & Bacon.

TEYBER, E. (1997). *Interpersonal process in psychotherapy*. Pacific Grove, CA: Brooks/Cole.

# 6

# Psychological Testing Overview

Counselors often hesitate to include psychological testing in their assessment of clients; some misunderstand the benefits of testing, and many regard testing with some skepticism. Their attitudes may be attributed to the traditional exclusionary use of testing to select those who will be admitted or rejected, to decide who will pass or fail, or to separate people into categories with labels. The historical concept and use of testing as a limiting or exclusionary tool may appear to be antithetical to the counseling profession's current orientation, which includes respecting and appreciating individual and cultural differences, viewing the client holistically, encouraging growth toward each client's optimal level of functioning, and empowering the client to take responsibility for self. Certainly, too, psychological testing has come under fire from various groups that are rightfully concerned about test bias and misuse of test results in our multicultural society (Hood & Johnson, 1991; Lewis, Hayes, & Lewis, 1986).

We would like to suggest that the counseling process can integrate psychological tests by using them in an inclusionary and adjunctive, rather than exclusionary, approach. That is, testing may be used as an additional important source of information within a comprehensive evaluation; this can assist both counselor and client in the following ways:

- Developing a more thorough understanding of the client
- Appreciating strengths, potentials, and limitations
- Identifying problem areas, including psychopathology

- Validating clinical impressions of our clients
- Making treatment plans and recommendations
- Setting goals and objectives for measurable treatment outcomes
- Evaluating client progress

We believe that if psychological testing is used sensitively and is respected as a component of the counseling process, it can be a therapeutic experience that helps to enhance the working alliance. In Chapter 5 we expressed a similar concept concerning the use of the initial intake assessment as an integral component of the counseling process and as an opportunity to build the therapeutic alliance.

We view psychological testing as a sort of second opinion about what may be going on in the client's life at the moment, instead of as a final, irrefutable fact. As with any opinion, which may vary over time and may be influenced by physical, emotional, and/or situational factors, the results of testing should always be taken with a grain of salt. (You may find it helpful to review your statistics notes for the concepts of reliability and validity at this point. You may also want to refer to Section E of the ACA Code of Ethics and Standards of Practice in Appendix E.) During your counseling internship and career, you may have access to modern computer scoring systems and methods that provide instant, lengthy, and often excellent "diagnostic" printouts based on the results of psychological testing. The trick here is to use modern technology without depersonalizing the client in the process. As counselors, we must value our professional knowledge and skills, our personal intuition, and our clinical interpretations as our most important therapeutic tools. View the client as a person, not a piece of interesting data!

Administering and interpreting psychological tests require advanced training, and in most settings, the responsibility for testing rests with licensed clinical, counseling, or school psychologists or licensed clinical mental health counselors. As a counselor intern, your exposure to psychological testing will vary greatly depending on your internship setting. At some agencies interns interface only infrequently with clinicians and clients who are involved in psychological testing, whereas at others interns are more actively engaged in the testing process. During your internship, a psychometrician or other professional who is competent in psychological testing should closely supervise your participation in this area. As a counselor intern in any setting, you should be familiar with the following basic areas of knowledge relevant to psychological testing:

- Human behavior, personality theory, physiological psychology, and psychopathology (in order to recognize a need for testing)
- Appraisal concepts, such as power versus speed, reliability, validity, standard deviation, mean
- Types of tests available—intelligence, personality, aptitude, achievement, interest inventory, and organicity

- Specific data concerning the particular instrument selected, such as norms, standardization criteria, cultural sensitivity, and applicability for individuals under the Americans with Disabilities Act

- Ethics and procedures related to the administration and interpretation of psychological tests

- General interpersonal and environmental conditions and individual situations that may influence test results

In addition, we as counselors and counselor trainees need to keep current with technical developments in testing so that we will be able to select the most appropriate instrument to provide information for our clients about their individual problems and concerns (Anastasi, 1992). Anastasi (1992) recommends that counselors take workshops and continuing education courses and read professional journals and other literature to stay abreast of the rapid advances in psychological testing. *Buros Mental Measurement Yearbook* (Conoley & Impala, 1998) and *Tests in Print IV* (Murphy, Conoley, & Impala, 1994) are valuable resources for counselors to familiarize themselves with current tests, their applicability to various client populations, and issues of reliability and validity.

We must also keep in mind that the testing experience itself affects the client and has the potential to be damaging as well as therapeutic. When clients are tested, results may be acutely influenced by clients' current interpersonal and environmental variables. Communication of test results is a particularly important aspect of testing, and we suggest that when you talk with clients about their tests, you try to:

- Help clients understand that the test is only one source of information and that the data gathered "suggests a hypothesis about the individual that will be confirmed or refuted as other facts are gathered" (Anastasi, 1988, p. 490).

- Relate test data to the client's functioning in real-life situations so that they are meaningful and helpful (Hood & Johnson, 1991).

- Emphasize that the test reflects the client's functioning in the past, but that this information may be used to make changes in the future (Hood & Johnson, 1991).

Many clients are anxious about tests and may need to be reassured concerning the purpose of the instrument and the information that will be elicited. As a counselor intern, you will need to attend carefully to your clients' feelings and thoughts about each aspect of the testing process. Taking a test evokes strong feelings and thoughts in most people, which provide fertile ground for exploration and the acquisition of new insight. One counselor intern told us that she had referred her client to the agency's consulting psychologist for an MMPI-2 to confirm a diagnostic impression of major depression and obsessive-compulsive disorder. The client was extremely upset afterward, saying, "That psychologist told me I am clinically depressed. How could he say that about me! I tried so hard on that test, too. I really thought I 'aced' it. I guess I can never do anything right." Another counselor intern learned never to test children during lunchtime, recess, or physical education class because their levels of motivation and interest may affect test results. An empathic response on the counselor's part helps to relieve anxiety, promote healing and growth, and build the therapeutic alliance. As an additional reference and solid foundation, we suggest counselors thoroughly review Section E of the American Counseling Association's Code of Ethics and Standards of Practice (1995), located in Appendix E.

Myriad psychological tests are extant (refer to Conoley & Impara, 1998, *The Thirteenth Mental Measurements Yearbook*, for a complete listing with descriptions). Some will have little value to you during your counseling internship, whereas others will offer relevant information for planning the best course of action for your clients. Just as tests are designed to categorize people and their problems, behaviors, style, or preferences, psychological tests themselves can be categorized into three discrete groupings: (1) tests of intelligence, (2) tests of ability, and (3) tests of personality (Anastasi, 1988). We suggest that you take the time, either before your internship or early in the experience, to review these general categories of tests and their uses and purpose.

Boughner, Hayes, Bubenzer and West (1994) surveyed members of the American Association for Marriage and Family Therapy and ascertained that they used 147 different standardized assessment instruments. They found that although respondents did not use standardized instruments often, the Myers-Briggs Type Indicator (MBTI) and Minnesota Multiphasic Personality Inventory-2 (MMPI-2) were most popular. Moreover, Bubenzer and his colleagues (1990) conducted a national survey of agency counselors to determine how often they used psychological tests. The following tests were used most frequently:

1. The Minnesota Multiphasic Personality Inventory (MMPI), now revised (MMPI-2) (Duckworth, 1991)

2. The Strong Campbell Interest Inventory (SCII), now revised Strong Interest Inventory (SII)

3. The Wechsler Adult Intelligence Scale-Revised (WAIS-R), now revised again (WAIS-III)

4. The Myers–Briggs Type Indicator (MBTI)

5. The Wechsler Intelligence Scale for Children-Revised, now revised again (WISC-III)

Our own experience confirms that these tests appear to be the most frequently used in clinical settings. We now discuss these five tests and how you, as a counselor intern, may use them to help your clients.

# THE MINNESOTA MULTIPHASIC
# PERSONALITY INVENTORY-2 (MMPI-2)

Lubin, Larsen, and Matarazzo (1984) determined that despite the fact that the original MMPI was the most widely used personality test in the United States (and perhaps in the world), critics expressed concerns about the instrument. For example, Graham (1990) notes archaic language and obsolete references in standardization samples and item content. Further, he cites sexist language and expressed some trepidation that the items had not been subjected to careful editorial scrutiny. Thus the MMPI was revised in 1989. The original MMPI was developed in 1943 by Hathaway and McKinley at the University of Minnesota Hospitals (Anastasi, 1988) as a tool to facilitate making differential diagnoses in a population of psychiatric inpatients. Graham (1990) notes that even from the outset, the MMPI was not very successful for this purpose and cannot separate people into distinct psychiatric categories. However, the MMPI has generated over 8,000 studies since it was published. The research has been so extensive and has provided so much information that the MMPI scales are successfully used to provide a wide range of data and inferences concerning an individual's personality, emotional characteristics, and psychological problems (Graham, 1990; Hood & Johnson, 1991).

The MMPI-2 contains 567 questions that the test taker answers as True, False, or Cannot Say. Butcher and Williams (1992) reported that elevated T-scores are 65 or greater (unless otherwise noted), whereas T-scores of 60 to 64 are moderately elevated. Generally, the more elevated the T-scores, the more likely the test taker is to have psychological or psychiatric problems. MMPI-2 results are interpreted, however, according to their "profile," overall configuration, and "code-type," or combination of the two or three most highly elevated scales' scores. An enormous amount of in-depth information about an individual's personality, symptoms, behaviors, and proclivities can be provided through analysis and interpretation of these code-types.

MMPI-2 scores are arranged in four validity scales that assess the subject's attitude toward the test and 10 clinical scales that assess emotional states, social attitudes, physical condition, and moral values (Graham, 1990). The validity scales include:

1. L, or Lie, which measures responses attempting to create a favorable impression, but which are likely untrue (faking good).

2. F, or Validity, which indicates carelessness, eccentricity, malingering (faking bad), confusion, or scoring errors.

3. ? , or Cannot Say, which indicates the number of questions left unanswered; this scale tends to invalidate the entire test if its score is too high.

4. K, or Correction, which indicates "faking good" or defensiveness if high and "faking bad" or self-denigrating attitude if low; this scale is used as a factor to adjust or correct other scale scores on the MMPI-2.

The 10 clinical scales include the following:

1. Hs: Hypochondriasis
2. D: Depression
3. Hy: Hysteria
4. Pd: Psychopathic deviate
5. Mf: Masculinity/femininity
6. Pa: Paranoia
7. Pt: Psychasthenia
8. Sc: Schizophrenia
9. Ma: Hypomania
10. Si: Social introversion

Numerous additional scales and subscales have been developed to augment and refine information obtained from MMPI-2 results; the most frequently used of these are generally included in computer-scored reports. The supplementary scales include the following:

- Anxiety (A) and Repression (R) Scales
- Ego Strength (Es) Scale
- MacAndrew Alcoholism-Revised Scale (MAC-R)
- Overcontrolled-Hostility (O-H) Scale
- Dominance (Do) Scale
- Social Responsibility (RE) Scale
- College Maladjustment (Mt) Scale
- Masculine Gender Role (Gm) and Feminine Gender Role (Gf) Scale
- Posttraumatic Stress Disorder (Pk) Scale
- Subtle-Obvious Subscales

Graham (1990) notes that examinees need answer only the first 370 items if just the standard scales are required for the assessment. Several researchers have also developed Critical Items Lists, which indicate areas warranting possible further evaluation by the mental health professional.

During your internship, the MMPI-2 may be useful to you in the following ways:

- You may have a source of information regarding your client's personality style or problems before your initial meeting with the client, so that you may modify your counseling interventions appropriately.
- You can use MMPI-2 results to confirm a diagnostic impression or to rule out a diagnosis. This may be especially helpful in determining whether to request a psychiatric consultation for psychotropic medication because

some disorders respond better than others to psychopharmacologic intervention.

- The test is valuable for augmenting information on clients who are not very verbal or have difficulty articulating their thoughts and feelings.

- You can discuss the critical items with your client as a way to explore problem areas and to help the client disclose important thoughts and feelings. Hood and Johnson (1991) suggest that counselors always discuss critical items with each client, especially those related to suicidal ideation. These authors write that clients may assume that the counselor is already aware of all their concerns because a question on the MMPI-2 alluded to an area, and may therefore not raise important issues during sessions.

- As with most other personality tests, results may change over time based on clients' emotional level of functioning. You can therefore compare scores from tests taken at different times to assess your client's response to life changes, stressors, or previous treatment, and you can also request an MMPI-2 after counseling to evaluate progress.

## THE STRONG INTEREST INVENTORY (SII)

The SII, formerly the Strong Campbell Interest Inventory (SCII), was revised in 1994. This inventory reflects clients' interests and capacities in the world of work and provides a dependable guide for career goals and development (Consulting Psychologists Press, 1994). It assesses individual likes, dislikes, and preferences for activities and also compares the similarities and differences of individual characteristics and preferences with those of a population already employed in a particular occupation. Many sources (Anastasi, 1988; Brown & Brooks, 1991; Hood & Johnson, 1991; Walsh & Betz, 1990) stress that interest inventories like the SII evaluate interests rather than abilities; therefore, they may be used only to predict how much an individual will enjoy an occupation or an academic area of study, rather than how likely he or she will be to succeed or how well he or she will do in a given field. The SCII was first published in 1974 and represented a non–gender-biased revision that combined two earlier versions of this instrument, the Strong Vocational Interest Blank for Men and the Strong Vocational Interest Blank for Women (Walsh & Betz, 1990).

Holland's theory, characterizing personality types and the type of work environment likely to be most congruent with each personality type, was the foundation for the structure of the SCII (Anastasi, 1988), and this continues to be the underlying SII basis. The instrument contains 317 questions, which yield data organized into the following categories:

- Four-color Profile with "snapshot" of results
- Six General Occupational Themes (RIASEC)
- 35 Basic Interest Scales

- Occupational Scales (109 occupations)
- Four Personal Style Scales
- Administrative Indexes

The Administrative Indexes provide information that may be used to help interpret the scale scores by addressing critical questions that must be answered in order to yield a confident interpretation, Infrequent Responses Scales indicating inappropriate response sets or respondent attitudes, Like/Indifferent/Dislike percentages, and the pattern of responses. The Basic Interest Scales report the client's interest in specific areas of activity, with higher scores indicating greater interest. The General Occupational Themes scales record the client's overall preferences related to work environment, activities, coping style, and types of people the client likes to be with, and yield a three-letter Holland personality code-type, such as RIA or ASE (Walsh & Betz, 1990). These personality types described by Holland (1985) include the following:

- Realistic (oriented toward tools or machinery)
- Investigative (oriented toward scientific endeavors)
- Artistic (oriented toward self-expressive activities)
- Social (oriented toward helping other people)
- Enterprising (oriented toward goal-directed business)
- Conventional (oriented toward clerical activity)

The Occupational Scales reflect similarity of the client's interests to the likes and dislikes expressed by people in various occupations. A higher score indicates that the client has more in common with people in a particular occupation. The Personal Style Scales note particular environments in which individuals like to learn and work as well as the type of activities they find rewarding.

Hood and Johnson (1991) stress that clients who are self-motivated by a desire for information will benefit most from an assessment process using an interest inventory. These authors also caution that the SII and other similar instruments are not suitable for use with clients who have emotional problems; these problems must be addressed before the client works on career or educational plans (Hood & Johnson, 1991). During your counseling internship, you may be able to use the SII to help your clients in the following ways:

- You can facilitate self-exploration and self-understanding as your client examines data regarding preferences, likes, and dislikes related to occupations, academic study, and other people.
- You can assist your client with educational plans related to career goals.
- You can provide new occupational options for your client to consider.
- You can help your client determine how much he or she has in common with other people who are employed in an occupational field.
- You can help validate your client's self-perceptions or clarify choices and thus build self-esteem.

# THE WECHSLER ADULT
# INTELLIGENCE SCALE-III

The Wechsler Adult Intelligence Scale, Third Edition, published in 1997, represents a revision of the Wechsler Adult Intelligence Scale-Revised. The WAIS Third Edition has the goals of establishing new norms within a more current representative sample, which includes expanded age ranges to reflect increased longevity (The Psychological Corporation, 1998). In addition, item content and item bias analyses were performed, artwork and materials were updated, and most subtests were altered specifically to assess individuals who have lower cognitive functioning. This instrument measures adult intelligence and is organized into a Verbal Scale, consisting of seven subtests, and a Performance Scale, consisting of seven subtests. These subtests are alternated, one from the verbal group and one from the performance group, as was the case with the WAIS-R. The scores yielded are a Verbal Score, a Performance Score, and a Full Scale score. The subtests are expressed as standard scores with a mean of 10 and a standard deviation of 3, whereas the Verbal, Performance, and Full Scale scores are reported as deviation IQs with a mean score of 100 and a standard deviation of 15.

The WAIS-III Verbal Scale consists of the following subtests:

- Information (questions are related to general, basic information, arranged according to level of difficulty, easiest to hardest)
- Digit Span (oral repetition of numbers, backward and forward)
- Vocabulary (words to be defined, arranged in order with easiest first and hardest last)
- Arithmetic (oral arithmetic problems)
- Comprehension (assesses understanding of meaning)
- Similarities (assesses ability to think in an abstract way)
- Letter-Number Sequencing (oral material that needs to be reordered and repeated): *supplementary subscale*

The WAIS-III Performance Scale consists of the following subtests:

- Picture Completion (cards in which the client must describe what is missing from each picture)
- Picture Arrangement (cards representing parts of a story that the client must arrange sequentially)
- Block Design (cards with colored patterns that the client must reproduce using colored blocks)
- Object Assembly (puzzles that must be assembled in a limited time): *optional subscale*
- Digit Symbol (a test of memory, eye-hand coordination, visual discrimination, and speed involving copying symbols for digits from a key card)

- Matrix Reasoning (examining geometric figures and either naming or pointing to the correct answer)
- Symbol Search (considering if target symbol groups appear within search symbol groups): *supplementary subscale*

The Wechsler Intelligence Scales may be used for differential diagnosis of psychological and mental problems, as well as for assessing intelligence (Anastasi, 1988; Hood & Johnson, 1991; Walsh & Betz, 1990). Variation among scores on subtests, differences among Verbal, Performance, and Full-scale scores, and overall patterns of subtest and scale scores are clinically significant. Careful analysis of Wechsler Intelligence Scales test results by a highly trained, experienced clinician can provide information that may indicate the possibility of drug or alcohol abuse, brain damage, Alzheimer's disease, anxiety, depression, reading problems, antisocial behavior, and other psychological dysfunctions (Anastasi, 1988; Hood & Johnson, 1991).

During your internship you may use the WAIS-III to help your client in the following ways:

- You may be able to modify or structure your counseling style and interventions to meet your client's needs more appropriately if you are aware of his or her intellectual functioning and cognitive style. For example, you may opt to use a more behavioral, less insight-oriented approach with a client who is developmentally challenged.
- You may help a client in career or educational planning.
- You may be able to help a client access beneficial community resources, such as special public school or housing programs to which he or she is entitled, if you are aware of intellectual, neurological, or learning problems.
- You may be able to use the test scores to facilitate treatment planning by confirming a diagnostic impression or ruling out a possible diagnosis.
- You may be able to help clarify your client's problem. For example, is the source of the difficulty a cognitive, neurological, or emotional problem?

## THE MYERS-BRIGGS TYPE INDICATOR (MBTI)

The most widely used personality instrument in the world today is the Myers-Briggs Type Indicator (Consulting Psychologists Press, 1998). Katherine Briggs and her daughter, Isabel Myers (Hood & Johnson, 1991) developed the Myers-Briggs Type Indicator in the 1920s, basing it on the work of Carl Jung. This instrument attempts to classify individuals into one of 16 specific personality types, using a four-letter code, and describes behavioral, emotional, and functional preferences. It contains 126 forced-choice questions. Hood and Johnson

(1991) note that the MBTI is popular and widely used because there are no good or bad or higher or lower scores, and no dysfunctions or diagnostic categories are implied by the test results. Rather, the instrument indicates tendencies toward certain characteristics and preferences but allows that individuals may also be able to use the opposite characteristic or preference.

The MBTI assesses the following personality factors:

1. Extroversion/introversion, an attitude describing the flow of psychic energy outward toward the world (E) or inward toward the self (I)

2. Sensing/intuitive, a perceiving function of receiving information from the outside world in a sensing (S) or intuitive (N) way

3. Thinking/feeling, a decision-making process based on data that have been received from the outside world in a thinking (T) or feeling (F) way

4. Judging/perceiving, the individual's dominant means of relating to the outside world in a judging (J) or perceiving (P) way

An example of a Myers-Briggs code-type might be ENFJ, meaning that the individual is an extroverted person, with a preference for absorbing and processing information in an intuitive and feeling way, who generally relates to the world with a judging attitude. The MBTI four-letter code attempts to present a comprehensive portrait of the way these dimensions of personality balance one another in an individual.

The instrument is relatively easy to score and interpret, and an instruction manual is included. The MBTI was the first widely used personality instrument developed and scored using the Item Response Theory (Consulting Psychologists Press, 1998). The Item Response Theory (IRT) method yields the best estimate of a person's best-fit personality type, the level of performance of each test item, and the selection of test items having the most meaningful measurement. Computer scoring methods also generate lengthy, detailed descriptions of client personality and behaviors. As always, we caution you to integrate test results as one data source within a comprehensive, evolving assessment of your client as a total human being.

The MBTI is appropriate for use in a variety of counseling settings because it provides abundant data in a nonthreatening, interesting way and because advanced training is not necessary for the mental health professional who scores and interprets this instrument. As a counselor intern, you may use the MBTI to help your clients in the following ways:

- You may use knowledge of your own code-type as well as your client's code-type to modify your counseling interventions and improve counselor–client communication and the working alliance.

- You can help your client gain self-understanding and explore important issues by becoming aware of how his or her ways of perceiving and relating to the world affect thoughts, feelings, and behaviors.

- You can help couples and families improve interpersonal relationships by gaining an understanding of how each of their code-types and personality styles may affect communication and attitudes.

- You can help your client choose work environments that are congruent with his or her code-type.

- You can help promote health and growth in your client by exploring the less-used functions of his or her personality style and code-type as potential coping mechanisms, new solutions to problems, alternative ways of relating to others, and new concepts of self.

## THE WECHSLER INTELLIGENCE SCALE FOR CHILDREN-III (WISC-III)

The Wechsler Intelligence Scale for Children was originally developed as a direct downward extension of an adult intelligence assessment instrument, the Wechsler-Bellevue Intelligence Scale (Seashore, Wesman, & Doppelt, 1950, cited in Hood & Johnson, 1991). First revised in 1974 (WISC-R), the present 1991 (WISC-III) revision maintains the basic structure and content of the WISC-R. In addition, WISC-III provides updated normative data, improved design and items, and three supplementary subtests (The Psychological Corporation, 1997). This revision continues to attempt to maintain balanced sensitivity to multicultural and gender references.

The WISC-III is similar in structure, content, and scoring organization to the WAIS-III (discussed earlier in this chapter). This test, which assesses individuals ages 6 through 16 years and 11 months, yields a Verbal Scale Score, a Performance Scale Score, and a Full-scale Score and contains 10 core and three supplementary subtests. The subtests are administered by alternating one from the Verbal scale and one from the Performance scale. The three supplementary subtests assess additional information about cognitive ability. The WISC-III subtests are also expressed as standard scores with a mean of 10 and a standard deviation of 3, whereas the Verbal, Performance, and Full-scale Scores are reported as deviation IQs with a mean of 100 and a standard deviation of 15.

The verbal scale of the WISC-III contains the following subtests:

- Information
- Similarities
- Arithmetic
- Vocabulary
- Comprehension
- Digit span (supplementary subtest)

The performance scale subtests are:

- Picture Completion
- Picture Arrangement
- Block Design
- Object Assembly

- Coding
- Symbol search (supplementary subtest)
- Mazes (supplementary subtest)

The WISC-III is scored, interpreted, and used in much the same way as the WAIS-III. This instrument can be used to assess children's intelligence, personality, and learning style, as well as for differential diagnosis of academic, neurological, and psychological problems. As a counselor intern, you may use the WISC-III to help your young clients in the following ways:

- You will be able to modify your counseling style and structure your interventions so that they will be more appropriate for your child client if you are aware of his or her cognitive and emotional developmental level.
- You may be able to clarify your client's difficulties as cognitive, neurological, emotional, or a combination of factors.
- You may be able to facilitate treatment planning by confirming a diagnostic impression or ruling out a possible diagnosis.
- You may be able to help parents or caretakers access beneficial school or community-based programs for which the child is eligible.
- You may be able to help parents and school personnel understand and meet the child's individual needs more effectively.
- You may be able to assist parents, caretakers, and school personnel in developing a realistic picture of the child's potential so that they can set appropriate goals.

## CONCLUDING REMARKS

Psychological testing can serve as one valuable source of data and an important component of a comprehensive assessment of the client. The testing experience can promote self-understanding and growth and can serve to strengthen the therapeutic alliance if you, as the counselor intern, attend to your client's feelings, thoughts, and behaviors throughout the testing process. Psychological testing should be viewed as a second opinion, a means of augmenting information and of increasing your understanding of the client's current concerns and current level of functioning. Results should be taken with a grain of salt! We believe that test results should be used to provide more pieces to the puzzle in understanding your client and should never be used as a means of disposition for any client.

## REFERENCES

ANASTASI, A. (1988). *Psychological testing* (6th ed.). New York: Macmillan.

ANASTASI, A. (1992). What counselors should know about the use and interpretation of psychological tests. *Journal*

*of Counseling and Development, 70*(5), 610–615.

BOUGHNER, S., HAYES, S., BUBENZER, D. & WEST, J. (1994) Use of standardized assessment instruments by marital and family therapists: A survey. *Journal of Marital and Family Therapy. 20* (1), 69–75.

BROWN, D., & BROOKS, L. (Eds.). (1991). *Career choice and development.* San Francisco: Jossey-Bass.

BUBENZER, D., ZIMPFER, D., & MAHRLE, C. (1990). Standardized individual appraisal in agency and private practice: A survey. *Journal of Mental Health Counseling, 12*(1), 51–66.

BUTCHER J. N., & WILLIAMS, C.L. (1992). *Essentials of MMPI-2 and MMPI-A interpretation.* Minneapolis, MN: University of Minnesota Press.

CONOLEY, J.C., & IMPARA, J.C. (Eds.). (1998). *The thirteenth mental measurement yearbook.* Lincoln, NE: The Buros Institute of Mental Measurements, The University of Nebraska.

CONSULTING PSYCHOLOGISTS PRESS, INC. (1994). *Strong interest inventory application and technical guide.* Palo Alto, CA: Consulting Psychologists Press, Inc.

CONSULTING PSYCHOLOGISTS PRESS, INC. (1998). *1998 preview: Myers-Briggs Type Indicator revision.* Palo Alto: CA: Consulting Psychologists Press, Inc.

DUCKWORTH, J. C. (1991). The Minnesota Multiphasic Personality Inventory-2: A review. *Journal of Counseling and Development, 69*(6), 564–567.

GRAHAM, J. (1990). *MMPI-2: Assessing*

*personality and psychopathology.* New York: Oxford University Press.

HOLLAND, J. (1985). *Making vocational choices: A theory of vocational personalities and work environments* (2nd ed.). Englewood Cliffs, NJ: Prentice-Hall.

HOOD, A., & JOHNSON, R. (1991). *Assessment in counseling: A guide to the use of psychological assessment procedures.* Alexandria, VA: American Counseling Association.

LEWIS, M., HAYES, R., & LEWIS, J. (1986). *The counseling profession.* Itasca, IL: Peacock.

LUBIN, B., LARSEN, R.M., & MATARAZZO, J.D. (1984) Patterns of psychological test usage in the United States: 1935–1982. *American Psychologist, 39*(4), 451–454.

MURPHY, L.L., CONOLEY, J.C., & IMPALA, J.C. (Eds.). (1994). *Tests in print IV.* Volume II. Lincoln, NE: The Buros Institute of Mental Measurements, The University of Nebraska.

THE PSYCHOLOGICAL CORPORATION (1997*). Wechsler intelligence scale for children* (3rd ed.). San Antonio, TX: The Psychological Corporation.

THE PSYCHOLOGICAL CORPORATION (1998). *Wechsler adult intelligence scale* (3rd ed.). San Antonio, TX: The Psychological Corporation.

WALSH, W., & BETZ, N. (1990). *Tests and assessments* (2nd ed.). Englewood Cliffs, NJ: Prentice-Hall.

# BIBLIOGRAPHY

COOPER, S. (1995). *The clinical use and interpretation of the Wechsler intelligence scale for children* (3rd ed.). San Antonio, TX: The Psychological Corporation.

GAY, L. (1987). Selection of measuring instruments. In L. Gay, *Educational research* (pp. 125–176). New York: Macmillan.

GRAHAM, J., SMITH, R., & SCHWARTZ, G. (1986). Stability of MMPI configurations for psychiatric inpatients. *Journal of Consulting and Clinical Psychology, 54*(3), 375–380.

HANSEN, J.C. (1992). *User's guide for the strong interest inventory* (rev. ed.). Stanford, CA: Stanford University Press.

KAUFMAN, A. (1994). *Intelligence testing with the WISC-III.* San Antonio: The Psychological Corporation.

LITWACK, L. (1986). Appraisal of the individual. In M. Lewis, R. Hayes, & J. Lewis (Eds.*), An introduction to the counseling profession* (pp. 251–277). Itasca, IL: Peacock.

OGDON, D. (1990). *Psychodiagnostics and personality assessment: A handbook.*

Los Angeles: Western Psychological Services.

THE PSYCHOLOGICAL CORPORA-TION (1997). *WAIS-III- WMS-III technical manual.* San Antonio: The Psychological Corporation.

WALLACE, S., & LEWIS, M. (1990). *Becoming a professional counselor.* New York: Sage.

WECHSLER, D. (1997). *The Wechsler adult intelligence scale* (3rd ed.)*: Administration and scoring manual.* San Antonio, TX: The Psychological Corporation.

# 7

# Integrating Psychotherapy and Pharmacotherapy

## Guidelines for Counselor Interns

The introduction of psychotropic medications during the early 1950s brought about a revolution in mental health care with the fortuitous discovery that an antihistamine, chlorpromazine (Thorazine), reduced the symptoms of psychosis in patients with schizophrenia (Klerman, 1991; Stahl, 1996). Over the past four decades, rapidly expanding research in neuroscience, biological psychiatry, and psychopharmacology has produced a profusion of new and highly effective psychotropic medications (Akiskal, 1989). These medications have decreased the frequency and length of psychiatric hospitalizations, alleviated the severity of symptoms in schizophrenia and other serious mental illness, provided hope and relief to millions of people suffering from depression and anxiety disorders, and reduced the stigma surrounding mental health treatment (Klerman, 1991; Yudofsky, Hales, & Ferguson, 1991). Pharmacotherapy has emerged as a predominant treatment modality because it is the least expensive and most rapid means of ameliorating mental and emotional disorders (Martindale, 1990).

## PSYCHOTROPIC MEDICATIONS: GOALS AND MECHANISM OF ACTION

Psychotropic medications can help restore individuals to an improved state of functioning by easing psychic pain, returning cognitive capacities, reducing social isolation, and replenishing hope and energy. However, these medications

**71**

are not intended to effect changes beyond the individual's normal capacities and personality (Schatzberg & Cole, 1991;Yudofsky, Hales, & Ferguson, 1991). Furthermore, because mental and emotional disorders always involve a dynamic interplay of biological, psychological, and social factors, psychotropic medications provide the greatest benefit when used in combination with psychosocial interventions and education.

Psychopharmacology is the study of how diseases affect the brain, how drugs impact the brain, and how these influences are expressed cognitively, affectively, and behaviorally. Principles of chemical neurotransmission provide the foundation for psychopharmacology (Stahl, 1996).

Neurotransmitters are the chemical messengers that direct and synchronize brain activity, consequently regulating thoughts, feelings, and behaviors. These messengers function by transmitting chemical and electrical messages across the neural synaptic spaces (the spaces between the nerve cells) in the brain. Multiple neurotransmitters may function simultaneously or consecutively in a single synaptic space. There are thought to be between several hundred and several thousand different kinds of naturally occurring neurotransmitters, and most psychotropic medications work by either altering or imitating the action of these neurotransmitters (Stahl, 1996).

## COMBINED TREATMENT: PSYCHOTHERAPY AND PHARMACOTHERAPY

Between 1970 and 1980, well-controlled double-blind outcome studies began to confirm the efficacy of combined pharmacological and psychotherapeutic treatment for mental and emotional disorders (Kahn, 1990). Research also suggests that combined treatment enhances patient compliance with psychotropic medications considered essential to treatment of certain conditions, such as mood stabilizers for bipolar disorder or antipsychotics for schizophrenia (Beitman, 1991; Paykel, 1995). Furthermore, evidence indicates that drugs can be prescribed at lower dosages when combined with psychotherapy than when used without concurrent psychotherapy (Paykel, 1995).

Due to a growing awareness of the complex and multicausal etiology of psychiatric disorders, the integration of psychotherapy and pharmacotherapy has become widely accepted practice. Combined treatment regimens are considered helpful in treating numerous mental and emotional disorders, including alcoholism, attention deficit–hyperactivity disorders, bipolar illness, chronic pain, dysthymic disorder, eating disorders, generalized anxiety, major depression, obsessive-compulsive disorder, panic, personality disorders, posttraumatic stress, premenstrual dysphoric disorder, schizophrenia and other psychotic processes, sexual dysfunction, sleep disorders, and tic disorders (Kaplan & Sadock, 1996; Maxmen & Ward, 1995a, 1995b; Pollack, Otto, & Rosenbaum, 1996; Reid, Balis, & Sutton, 1997).

Psychotropic drugs comprise a large proportion of the most commonly prescribed medications; they are often prescribed not only by psychiatrists but also by health care providers specializing in areas such as family practice, pediatrics, neurology, osteopathy, and dentistry. Because psychotropic medications are so prevalent, particularly in the clinical mental health field, counselors and counselor interns need to develop familiarity with them, as well as skills in interfacing with individuals who are taking, or who may need referral for, psychotropic medications.

## THE ROLE OF COUNSELORS IN COMBINED TREATMENT

As a counselor intern, you may be the mental health professional with whom the client interacts most often. Your responsibilities may include recognizing when a referral for medication evaluation is appropriate, making referrals to psychiatrists, encouraging and monitoring medication compliance, inquiring about side effects and symptoms, educating clients and their families about medication, and communicating with physicians about the client's status. Because these professional tasks require highly specialized knowledge and clinical skills, you should always consult carefully with your site supervisor if you are concerned about a client's potential need for medication referral.

The following sections of this chapter present guidelines for counselor interns working with clients who are taking psychotropic medications, as well as those who may need referral to a psychiatrist for medication evaluation.

## CLIENTS' FEELINGS ABOUT MEDICATION

Martindale (1990) presents an important clinical maxim succinctly: "Even in the most compliant of patients, the taking of medication is not an emotionally neutral act" (p. 335). Although pharmacological intervention emerged from a purely biological model, it encompasses many of the same issues and dynamics as do other psychotherapeutic treatment modalities (Gabbard, 1990). Counselors need to attend carefully and empathically to their clients' concerns about medications.

### Negative Feelings

Referral for medication can elicit negative feelings in clients. A combined treatment situation involves a physician in addition to the counselor and the client, thus transforming the therapeutic dyad into a therapeutic triad. This change alone adds complications that can potentially undermine the original therapeutic alliance (Pilette, 1988). Clients referred for medication may feel as though they are being rejected because the counselor dislikes them or considers them uninteresting or because they are so ill that the counselor has given up on them (Chiles, Carlin, Benjamin, & Beitman, 1991; Bradley, 1990; Busch & Gould, 1993). Many clients interpret the referral as punishment for their failure to work hard enough in psychotherapy or for their lack of psychological mindedness. Sometimes clients fear that their families will view medication as confirmation that the client is severely mentally ill or is the cause of all family problems.

All negative feelings need to be explored with sensitivity and respect and addressed actively as issues about treatment. In combined treatment, medication compliance is a primary treatment objective, and deep exploration of feelings about taking medication is not necessarily required to enhance compliance. The rationale for combined psychotherapy and pharmacotherapy should be re-stated clearly to the client. The clinician hoping to enhance compliance should never react impatiently to the client's feelings or worries about medications.

## Positive Feelings

Medication referral can elicit positive feelings because clients may experience the referral as acknowledgement of their psychic pain or as a nurturing response from their therapist (Bradley, 1990). Bradley (1990) writes that some clients may be more satisfied that they are receiving excellent treatment when two specialists are involved, and others may even feel that they have achieved the status of a favorite child because their two therapists confer about them.

## Splitting

Medication referral may provoke defensive splitting by the client because the psychotherapist may be viewed as incompetent or unable to provide help without assistance from another professional. The psychiatrist, on the other hand, may be cast as omnipotent and benevolent. Splitting is often reinforced by the characteristically different therapeutic approaches of psychotherapists and pharmacotherapists, as well as by the different mechanism of action of each treatment modality (Busch & Gould, 1993). Typically the psychotherapist is more nondirective, withholding advice and maintaining a neutral stance, while the pharmacotherapist is more directive, giving advice and assuming a persuasive or coercive position (Bradley, 1990). Psychotherapy offers gradual improvement and entails discomfort and hard work, while medication offers rapid symptom resolution without effort on the part of the client (Busch & Gould).

Splitting should be discouraged because it is detrimental to treatment (Bradley, 1990; Busch & Gould, 1993; Chiles, Carlin, Benjamin, & Beitman, 1991). Both clinicians should be attuned to the potential for splitting in the three-way treatment relationship, and both must attend carefully to their own counter-transference feelings. For example, the clinicians may act out the client's split countertransferentially when the psychiatrist feels annoyed at having to share control of the case with the counselor, and the counselor feels devalued by the power of the psychiatrist's medication (Bradley, 1990; Busch & Gould, 1993). Chiles, Carlin, Benjamin, & Beitman (1991) suggest that frequent contact between clinicians ensures a joint approach toward concerns such as treatment goals, relationship issues, and medication compliance; this also provides a comforting sense of consistency for the client (p. 114).

## Fears and Worries

Sometimes clients fear that taking medication demonstrates weakness of character or a propensity for dependency. In this case the clinician can offer the following therapeutic response: "If you have a broken leg, you use a cast for support until the bone heals. Using a cast is not a sign of being weak or dependent. It is the right way to treat a broken bone. The medication is like a cast—it provides support until you heal, and it is the right way to treat this problem. Medication is not a sign of being weak or dependent."

Clients who require a maintenance regime frequently protest that they do not need medication any longer because they are currently symptom-free. In this situation a statement such as the following may be helpful: "People who have diabetes remain healthy as long as they take medicine to control it. When they are taking their medicine, they may even feel and look so healthy that it is easy to forget they have diabetes. But if they stop taking their medicine, they soon become very ill. The same is true for people with bipolar illness (or schizophrenia, recurrent major depression, and so on). Taking the medicine regularly keeps you healthy."

Clients often forget which symptoms were present before starting drug therapy and incorrectly attribute these symptoms to their medication. For example, they may complain that the medicine is keeping them awake at night or making it hard to concentrate and then insist that they must stop the medication because they cannot tolerate these side effects. Clinical wisdom dictates documenting symptoms throughout treatment and sharing this information with clients who need reassurance that particular symptoms are manifestations of the disorder rather than side effects of the medication.

## Multicultural Issues

Both the counselor and the physician need to be sensitive to multicultural values and symbolic meanings that may impact treatment. Some cultures regard medication as integral to treatment, while others perceive medication as a sign of personal weakness or failure. For example, many Hispanic cultures embrace a universal expectation for the helping professional to provide a tangible mechanism of treatment, such as medication (Ward, 1991). In contrast, referral for medication may be interpreted as a threat in cultures valuing independence and self-control, or as an affront in cultures holding psychological insight and self-awareness in high esteem. Intense cultural stigmatization surrounding mental illness may be heightened by the clinician's suggestion that psychotropic medication is indicated.

## Client Requests for Medication

Occasionally clients request medication referral, and although the request may be appropriate, it may also have latent meaning or emotional significance or may represent cognitive distortions. Moreover, in our "take a pill" society, some clients may expect to find an instant cure for their problems. Frequently clients who ask for medication are acquiescing to pressure from family members or friends who may be trying to help or need to feel important or to control the client. Sometimes a request for medication expresses the client's yearning for more affection from the counselor or reflects a developmental need for a concrete representation of the counselor's care. In fact, some clients come to depend on their medications as transitional objects. The counselor should also be alert to a request for medication as a sign that the client is feeling overwhelmed with painful affect, perhaps because the pace of psychotherapy is too rapid. Finally, some clients' demands for medication represent their characteristic defensive strategy of using substances to numb feelings.

## TALKING TO CLIENTS ABOUT MEDICATION

In combined treatment, the protocol for both pharmacotherapy and psychotherapy sessions should include an inquiry about progress made or problems encountered in the complementary component of the treatment. The counselor in a combined treatment arrangement therefore needs to provide more structure for the session than might ordinarily be required in psychotherapy alone. The agenda for every session should include questions pertaining to medication, including a review of symptoms and questions about compliance and side effects (Wright & Thase, 1992).

### Initiating the Discussion About Medication Referral

Chiles, Carlin, Benjamin, & Beitman (1991) suggest that, upon initiating a discussion about referral for medication, the referring clinician should present the client with a clear rationale for combined psychotherapy and pharmacotherapy. It is often helpful for the counselor to illuminate the salience of both psychological and biological factors to the client's problem, explaining that medication may reduce uncomfortable symptoms while psychotherapy teaches problem-solving skills and new behaviors (Chiles, Carlin, Benjamin, & Beitman, 1991).

Many clients respond well to a statement such as "The medication can take the edge off your pain so that you will be more free to work in psychotherapy, learning how to change things that may have contributed to your difficulties." The statement should be tailored to the specific needs and situation of the client. Here are some examples: "The medication can help calm some of your anxiety so you will be more able to relax in psychotherapy and learn how to change things that may have contributed to your difficulties," or "The medication can relieve some of your racing thoughts so you will be more able to concentrate in psychotherapy and learn how to change things that may have contributed to your difficulties."

## Delineating Professional Roles

A model of combined treatment, as well as the role of each professional and his or her responsibilities to the client, should be clearly delineated in the initial discussion about referral. Clients sometimes assume that a referral for medication means the counselor intends to relinquish care to the psychiatrist. Unhappy misunderstandings can be avoided by reiterating to clients that both professionals will continue to participate in the treatment. For example, the counselor might say, "Dr. Smith will see you to decide whether medication might be helpful and to select one to prescribe. He will most likely see you several times at first to make sure things are going smoothly, but eventually he will see you only once every few months. Your visits to Dr. Smith will not change the frequency of your meetings with me. I will still continue to see you according to our regular therapy schedule even when you are taking medications."

## Tailoring Discussions to the Client

Several authors (Bradley, 1990; Chiles, Carlin, Benjamin, & Beitman, 1991; Ward, 1991) stress the importance of attending carefully to the client's personality type, cognitive style, needs for interpersonal closeness or distance, and personal conceptualization of symptoms when discussing referral for medication. For example, the client with a dependent or histrionic personality style may do well with a warm approach but only a general explanation about the potential benefit of medication, while a client with a paranoid or obsessive-compulsive style might respond better to a more formal discussion presenting a great deal of detailed information (Chiles, Carlin, Benjamin, & Beitman, 1991).

Anxious, suspicious, or obsessive-compulsive clients may fear change and may need reassurance that the medication may result in small improvements in mood and thinking but cannot cause major alterations in personality (Chiles, Carlin, Benjamin, & Beitman, 1991; Ward, 1991). Statements such as the following may be helpful: "The medication can help you feel more like you did before you began having these problems, but it certainly cannot change your personality. You will still be exactly the same person, although you may feel a little bit less anxious."

As Ward (1991) writes, help-rejecting clients who habitually comment, "I've already tried something like that and it didn't work," sometimes respond well to a paradoxical intervention from the clinician: "This may not work for you either. In fact, I have real doubts about much improvement, and you will probably have some side effects. Do you still think you want to give it a try?" According to Ward, these clients are determined, at some level, to prove the clinician wrong, and may therefore allow themselves to be helped by medication only if the clinician predicts that there won't be much improvement.

Often using clients' own conceptual models of their problems may encourage receptivity to a medication referral. For example, clients who consider the need to take medication for an emotional difficulty to be a sign of weakness may be quite relieved to accept a referral for medication that can help with stress, insomnia, or low energy. Similarly, an acutely manic client talking about "cosmic consciousness" may refuse the counselor's referral for medication to reduce mood swings but may accept a referral for medication to help with "centering" or "focusing" (Ward, 1991, p. 81).

Ward (1991) suggests that the clinician must always convey an understanding and appreciation of the client's own belief system. For example, the client who wants to use natural or alternative remedies should be supported as long as these remedies are not harmful. Once the advantage of a holistic approach to health has been validated, medication can be discussed as an adjunctive treatment to help restore the body's natural balance (Ward, 1991).

## General Therapeutic Statements

Most clients feel encouraged when the counselor mentions the following:

1.  Starting medication is in some ways like buying a new pair of shoes. You often need to try on several different pairs in order to find the one just right for you. Sometimes several different medications must be tried, as well, to find the one just right for you—the one that helps you the most and has the fewest side effects.

2.  Most medications require two weeks or more to reach therapeutic levels in the blood until they can begin to help you feel better.

3.  Sometimes side effects or medication-related symptoms, such as stomach upset, are prominent when a drug is first started but then gradually subside or disappear with time. Patience helps.

4.  Taking a medication definitely does not mean you are crazy. In fact, making good use of all available resources is a sign of good judgment.

5.  The purpose of medication is to help you feel more like yourself at your best. Medication cannot make you "high" or change your personality.

## WHEN TO REFER

As a counselor intern, you should consider referral to a psychiatrist for medication evaluation whenever you recognize a constellation of symptoms known to be amenable to pharmacological intervention or whenever symptoms have not responded adequately to psychotherapy (Busch & Gould, 1993). When in doubt, always seek advice from your site supervisor. Bradley (1990) emphasizes that medication referral is not an appropriate means of overcoming a therapeutic impasse, obtaining professional supervision, or withdrawing from care of a difficult client.

The following clinical situations are among those that warrant referral for medication consultation:

1.  The presenting problem is a recurrence of, or similar to, a past problem that responded favorably to treatment with psychotropic medications.

2. The client presents with significant suicidality or a history of suicide gestures or attempts.

3. There are significant symptoms of depression, including changes in appetite and weight, sleep disturbance, low energy, difficulty concentrating, hopeless feelings, psychomotor retardation, extreme fatigue, and thoughts of death. In children, depression may be expressed as irritability or angry behavior, or the child may talk about wanting to die or may produce drawings or notes about death.

4. The client appears to be markedly confused or complains of memory loss or other cognitive impairment.

5. There are signs of a manic episode, including abnormally elevated or irritable mood, pressured speech, greatly increased energy, psychomotor agitation, reduced sleep, and excessive involvement in activities with a high potential for destructive consequences (such as sudden high frequency of unprotected sexual activity with multiple partners, unusual number of costly items purchased, or large amount of money spent quickly). In children, symptoms of mania may be expressed as irritability, angry or out-of-control behavior, or great difficulty settling down.

6. The client presents with symptoms of psychosis such as hallucinations, delusions, or otherwise impaired reality testing.

7. The client complains of obsessive thoughts, compulsive behaviors, or both.

8. The client complains of panic or unremitting anxiety.

9. The client's complaints are largely somatic, such as headaches, dizziness, numbness, or weakness.

10. There is a history of significant mood swings or a suspicion of mood swings in cases where a family history of bipolar illness exists.

Additional clinical presentations certainly support referral to a psychiatrist; however, the scope of this chapter is limited to referral for medication evaluation.

## PSYCHOTROPIC MEDICATIONS

The rapid pace of research in the neurosciences, biological psychiatry, and psychopharmacology ensures that any review of psychotropic medications is likely to be outdated shortly after its publication. Counselors involved in combined psychotherapy and pharmacotherapy must stay current with the rapidly changing literature in the field. With this caveat, the following sections of this chapter offer a brief summary of five major classes of psychotropic medications: antidepressants, antianxiety agents, mood stabilizers, antipsychotics, and psychostimulants. For more detailed information, counselors are encouraged to refer to the most recent edition of the *Physician's Desk Reference* (*PDR:* Medical Economics Company, 1998), which provides a comprehensive and reliable resource, or to other current psychopharmacology texts.

# ANTIDEPRESSANTS

Antidepressants reduce the affective, cognitive, and behavioral symptoms associated with major depression, dysthymia, atypical depression, and seasonal affective disorder, such as dysphoric mood, sleep and appetite disturbances, anhedonia, low energy and fatigue, feelings of guilt, rumination, social withdrawal, hopelessness, helplessness, and thoughts of death. Antidepressants have also recently been approved as a primary treatment for obsessive-compulsive disorder and may sometimes be used to treat the following, without formal FDA approval in all cases: addictions, anxiety, attention-deficit disorders, borderline personality disorder, childhood enuresis, chronic pain, eating disorders, insomnia, migraines, panic, premenstrual dysphoric disorder, posttraumatic stress, premature ejaculation, schizoaffective disorder, and social phobia (Kaplan & Sadock, 1996; Maxmen & Ward, 1995a, 1995b; Pollack, Otto, & Rosenbaum, 1996; Reid, Balis, & Sutton, 1997).

Clients presenting with symptoms of depression who have not had a recent physical exam should be referred to their physicians for a check-up, because depressed mood can be associated with a range of physical conditions. These include brain injury, diabetes, Huntington's disease, Lyme disease, multiple sclerosis, pancreatic cancer, parathyroid disease, Parkinson's disease, pellagra, porphyria, stroke, syphilis, thyroid disease, and Wilson's disease (Morrison, 1997).

Antidepressants have little or no potential for promoting drug dependency, addiction, or tolerance. Significant concerns in using these medications are their troublesome side effects, potential for serious drug or food interaction that may occur with certain types of antidepressants, and lethality of overdose. Antidepressants may be prescribed in combination with other drugs to enhance their therapeutic potency (Yudofsky, Hales, & Ferguson, 1991). Medications used to augment antidepressants include low doses of lithium, certain stimulants, and commercially manufactured thyroid hormone (Schatzberg & Cole, 1991).

On average, the duration of an untreated episode of depression is 6 to 24 months (Stahl, 1996), whereas the duration of a treated episode is about 3 months (Wolski, 1995). Treatment with antidepressants for three to six weeks is usually required before full therapeutic effects are seen, and many treatment failures are actually due to insufficient trial of the medication. One new medication, Celexa (citalopram), introduced in the United States in 1998, reportedly decreases depressive symptoms within one week of initiating treatment.

Depression is a recurrent illness, and a biological mechanism known as *kindling* results in structural and biochemical changes in the brain that increase the severity of each subsequent episode and decrease the time between episodes following the initial episode of depression. Furthermore, recurrence occurs in 50 percent of clients having one episode of depression, 70 percent of those having two episodes, and 90 percent of those having three episodes (Stahl, 1996). Therefore, as a rule, clients are treated with antidepressants for at least six months after the first episode to reduce incidence of relapse or recurrence. Stahl notes that although there are no formal guidelines for treating clients who have suffered multiple depressive episodes, many psychiatrists consider it prudent to maintain these clients on antidepressants indefinitely as a preventive measure.

Sharing information about the predicted course of recovery on antidepressant medications communicates security to clients and helps them maintain a reassuring sense of control because they know what to expect. The expected sequence of symptom improvement for individuals treated with antidepressants is as follows: (a) insomnia diminishes in 3 to 4 days; (b) appetite returns in 5 to 7 days; (c) energy increases in 4 to 7 days; (d) mood, concentration, and interest in activities begin to improve in 7 to 10 days; (e) libido returns in 9 to 10 days; (f) hopelessness and helplessness decrease in 10 to 14 days; (g) ability to enjoy activities begins to return in 10 to 14 days; (h) sadness, guilt, and suicidal ideation decrease in 12 to 16 days (Maxmen & Ward, 1995b).

There are four classes of antidepressants: atypical antidepressants, serotonin specific reuptake inhibitors (SSRIs), heterocyclics, and monoamine oxidase inhibitors (MAOIs). The atypical antidepressants and the SSRIs are the most recently introduced and widely prescribed antidepressants. They are generally associated with fewer adverse side effects and lower risk of lethal overdose than the MAOIs and the heterocyclics. In certain cases, heterocyclics and SSRIs can be used to augment each other (Maxmen & Ward, 1995b).

## Atypical Antidepressants

The side effect profiles and therapeutic benefits of the atypical antidepressants vary considerably. The compliance rate for this class of medications is high overall because side effects are generally mild.

Table 7.1 on the following page lists trade and generic names, side effects, and other information for the atypical antidepressants.

## Serotonin Specific Reuptake Inhibitors (SSRIs)

The SSRIs produce a broad spectrum of therapeutic benefits. Their mechanism of action is thought to be the inhibition of absorption of serotonin, the body's natural mood-enhancing chemical, so that it remains in the synaptic spaces of the brain for an extended time. All the SSRIs share class-specific side effects, with minor variations, depending on the particular medication and the individual client. In general, common side effects include anxiety, dizziness, gastrointestinal upset, insomnia, and sexual dysfunction. Less common side effects include fatigue, somnolence, sweating, tremor, and unusual or vivid dreams. Risk of lethal overdose is extremely low. Fatal drug interactions between some SSRIs and MAOIs have been reported, so these medications should never be used together.

Table 7.2 on page 83 lists trade and generic names, side effects, and other information for the SSRIs.

**Table 7.1  Atypical Antidepressants**

| Trade Name | Generic Name | Side Effects | Other Facts |
|---|---|---|---|
| Desyrel | Trazodone | Dizziness, dry mouth, postural hypotension, sedation | Should be taken with food or at bedtime. Used in small doses to treat insomnia. *Men should always be warned of possibility of sustained erection (priapism) requiring emergency medical intervention.* |
| Effexor | Venlafaxine | Appetite changes, constipation, dizziness, dry mouth, G.I. upset, headache, insomnia, sedation, somnolence, sweating, tachycardia | Highly effective treatment for melancholic depression; safe for geriatric clients |
| Serzone | Nefazodone | Dizziness, dry mouth, headache, sedation, somnolence | Does not cause anxiety or insomnia. Low risk of mania induction in bipolar illness. |
| Wellbutrin | Bupropion | Agitation, anxiety, appetite/weight changes, constipation, insomnia, sweating | Highest compliance rate of antidepressants. Increased risk of seizure at high doses (>400mg.), with initial rapid increase in dosage, or in clients having bulimia or seizure disorders. New sustained release formulation reduces risk of seizures. |

## Heterocyclics

Approximately 80 percent of people with major depression improve when treated with heterocyclics (Yudofsky, Hales, & Ferguson, 1991). However, the potential of lethal overdose is very high with this class of antidepressants, and because the target population is frequently at risk for suicide, psychiatrists often prescribe only small amounts of heterocyclics at one time. Nevertheless, clients sometimes hoard pills until they have accumulated enough for a fatal overdose, and therefore careful monitoring of suicidal ideation and feelings of worthlessness or hopelessness is mandatory.

Class-specific side effects are particularly problematic with heterocyclics, although the severity of each of the side effects varies tremendously with particular medications (Maxmen & Ward, 1995b). *Anticholinergic* side effects generally include blurred vision, confusion, constipation, dizziness, dry eyes, dry mouth and throat, memory problems, nasal congestion, near-angle glaucoma, hypersensitivity to light, slowed G.I. function, and urinary problems. Many of these problems resolve with time or may be treated symptomatically, such as by directing clients to sip beverages or suck on hard candies to relieve dry mouth or to increase dietary bulk to reduce constipation.

## Table 7.2 Serotonin Specific Reuptake Inhibitors (SSRIs)

| Trade Name | Generic Name | Side Effects | Other Facts |
|---|---|---|---|
| Celexa | Citalopram | Dry mouth, G.I. upset, insomnia, somnolence | Newest SSRI; released in 1998. Therapeutic effects occur within one week of starting Celexa. Not associated with sexual dysfunction. Dangerous drug interactions with MAOIs. |
| Luvox | Fluvoxamine | Anxiety, appetite decrease, dry mouth, G.I. upset, insomnia, sedation, tremor | Approved treatment for OCD. Dangerous drug interactions may occur with alprazolam (Xanax), astemizol (Hismanol), and triazolam (Halcion). Abrupt withdrawal may result in headache, dizziness, nausea. |
| Paxil | Paroxetine | Constipation, dizziness, dry mouth, G.I. upset, headaches, sedation, sexual dysfunction, sweating | Approved treatment for OCD. Flu-like syndrome may occur when Paxil is discontinued unless dose is tapered gradually. |
| Prozac | Fluoxetine | Anxiety, constipation, G.I. upset, headaches, insomnia, sedation, sexual dysfunction, tremor, weight loss | Approved treatment for OCD. Best taken in morning due to energizing effect. Longest half-life of SSRIs. Dangerous drug interactions with MAOIs; wait 5 weeks after stopping Prozac before taking MAOIs. No research evidence to support increased suicidality, despite media reports (Stahl, 1996). |
| Remeron | Mirtazapine | Appetite increase with weight gain, dizziness, elevated cholesterol and triglycerides, sedation, somnolence | Dangerous drug interactions with MAOIs; wait 14 days after stopping Remeron before taking MAOIs. |
| Zoloft | Sertraline | Dizziness, dry mouth, ejaculatory delay in males, G.I. upset, insomnia, somnolence, sweating, tremor | Dangerous drug interactions with MAOIs; wait 14 days after stopping Zoloft before taking MAOIs. Abrupt withdrawal may result in headache, dizziness, nausea. |

### Table 7.3 Heterocyclics

| Trade Name | Generic Name | Side Effects | Other Facts |
|---|---|---|---|
| Adapin | Doxepin | Blurred vision, constipation, dizziness, dry mouth, postural hypotension, sweating, sedation, tachycardia, weight gain | |
| Elavil, Endep | Amitriptilyline | Blurred vision, confusion, constipation, dizziness, dry mouth, EKG abnormalities, fatigue, hypertension (blood pressure), postural hypotension, sedation, sweating, tremor, weakness, weight gain | |
| Anafranil | Clomipramine | Anxiety, appetite changes, blurred vision, constipation, dizziness, EKG abnormalities, fatigue, G.I. upset, menstrual changes, muscle cramps, postural hypotension, sedation, sexual dysfunction, sore throat, sweating, tachycardia, tremor, weakness, weight gain | Highly effective treatment for OCD |
| Asendin | Amoxapine | Constipation, dizziness, dry mouth, insomnia, postural hypotension, sedation, skin rash, tachycardia, urinary hesitancy or retention | |
| Ludiomil | Maprotaline | Blurred vision, dry mouth, constipation, sedation, skin rash, weight gain | |
| Norpramin | Desipramine | Dry mouth, skin rash | |
| Pamelor | Nortriptyline | Confusion, dry mouth, fatigue, tremor | |
| Sinequan | Doxepin | Sweating, sedation, tachycardia, weight gain | |
| Surmountil, Vivactil | Trimipramine | Confusion, constipation, dizziness, dry mouth, EKG abnormalities, postural hypotension, sedation, tremor, weight gain | |
| Tofranil | Imipramine | Agitation, anxiety, dizziness, EKG abnormalities, excitement, headaches, hypomania, insomnia, sedation, sweating, tachycardia, weakness, weight gain | May cause increased risk of falls in elderly. |

Table 7.3 lists trade and generic names, side effects, and other information for the heterocyclics.

## Monoamine Oxidase Inhibitors (MAOIs)

The MAOIs are especially effective in treating atypical depression, depression with a large component of anxiety, and treatment-resistant mood disorders. Because risk of lethal overdose with MAOIs is exceedingly high, clients taking these medications must be closely monitored for suicidality (Maxmen & Ward, 1995b). Class-specific side effects include postural hypotension, weight gain, insomnia, and sexual dysfunction.

Most physicians advise that clients discontinue MAOIs before undergoing surgery because of possible blood pressure complications (Schatzberg & Cole, 1991). Another major concern in using this class of antidepressant is the severe, often fatal interactions with both food and drugs that MAOIs can precipitate. Individuals taking MAOIs must avoid using several types of prescription and over-the-counter medications and must restrict their diets carefully to avoid foods high in tyramine. Therefore, MAOIs should not be prescribed for clients who are confused or who have memory or concentration problems.

The following is a partial list of foods that clients should avoid when taking MAOIs:

- Alcoholic beverages (beer, champagne, hard liquor, chianti wine, sherry)
- Aspartame (artificial sweetener)
- Cheese (including blue, brie, cheddar, Gruyere, Swiss, and others)
- Chinese food (including soy sauce and miso)
- Delicatessen specialty meats and fish (such as salami, sausage, mortadella, pickled herring)
- Fruits (figs, overripe bananas, raspberries)
- Sauerkraut
- Yeast

The following is a partial list of drugs that clients must avoid when taking MAOIs:

- Antihistamines (such as Chlor-Trimeton)
- Appetite suppressants
- Asthma medications (such as Proventil)
- Amphetamines (such as Dexedrine)
- Decongestants and cough suppressants (such as Robitussin)
- Demerol
- Local anesthetics (such as Novocain)
- Phenylpropanolamine (such as Alka-Seltzer Plus or Allerest)
- Pseudoephedrine (such as Actifed, Sudafed, Sinutab, or Vick's NyQuil)
- Ritalin
- SSRIs
- Heterocyclics

**Table 7.4 Monoamine Oxidase Inhibitors (MAOIs)**

| Trade Name | Generic Name | Side Effects | Other Facts |
|---|---|---|---|
| Nardil | Phenelzine | Blurred vision, dizziness, dry mouth, edema, excitement, G.I. upset, hypomania, postural hypotension, insomnia, sedation, sexual dysfunction, tachycardia, tremor, urinary hesitancy or retention, weight gain | Abrupt discontinuation may cause agitation, anxiety, nightmares, psychosis. |
| Parnate | Tranylcypromine | Dizziness, dry mouth, constipation, excitement, G.I. upset, headaches, hypomania, insomnia, postural hypotension, sedation, tachycardia | Abrupt discontinuation may cause agitation, anxiety, confusion, depression, delirium, headaches, insomnia, muscle weakness, nightmares, psychosis. Small potential for tolerance and addiction. |

Clients using MAOIs should be careful to keep them out of reach of children and should carry a Medic Alert card informing medical personnel about MAOI use (Kaplan & Sadock, 1996).

Table 7.4 lists trade and generic names, side effects, and other information for the MAOIs.

## ANTIANXIETY AGENTS (ANXIOLYTICS)

Anxiety is a normal, adaptive human reaction to threat that evolved as a component of the fight or flight survival response (Stahl, 1996). However, anxiety is considered maladaptive when it impedes functioning or is aroused by persons or situations that do not pose any real danger. Anxiety that interferes with daily activities has been identified as "the most widespread of all psychological problems" (Yudofsky, Hales, & Ferguson, 1991, p.71).

Antianxiety agents are used to alleviate the physical and psychological symptoms that may be caused by anxiety, including chills, clammy hands, difficulty concentrating, difficulty swallowing, dizziness, dry mouth, fatigue, fearfulness, frequent urination, gastrointestinal upset, headaches, insomnia, irritability, muscle tension and aches, sweating, tachycardia, and tremor. Clients presenting with symptoms of anxiety should be referred to their physicians for an examination to rule out a physical cause for their problems. Anxiety can be associated with the following physical conditions: premenstrual dysphoric disorder, cardiac arrhythmias, chronic obstructive lung disease, lung cancer, mitral valve prolapse, and lung disease (Morrison, 1997).

Antianxiety agents are used to treat acute stress reactions, adjustment disorder with anxious mood, agitated depression, agoraphobia, alcohol withdrawal, elective mutism, generalized anxiety disorder, insomnia, night terrors, panic disorder, performance anxiety, phobias, posttraumatic stress, obsessive–compulsive disorder, and rage attacks (Kaplan & Sadock, 1996; Maxmen & Ward, 1995a, 1995b; Pollack, Otto, & Rosenbaum, 1996; Reid, Balis, & Sutton, 1997). Antianxiety agents are also used as muscle relaxants, to control seizures, as preanesthetic agents for surgery, to treat migraines, and as short-term treatment for some of the side effects associated with SSRIs and atypical antidepressants.

There are five classes of antianxiety agents: azaspirone, barbiturates, benzodiazepines, antihistamines, and beta-blockers.

## Azaspirone

Buspar (buspirone) is the only medication in the azaspirone class. Introduced within the last several years, it provides safe and effective treatment for anxiety. Buspar is less sedating than the other antianxiety agents, and does not impair coordination or balance. It does not interact adversely with alcohol, nor does it promote addiction, dependency, or tolerance. However, treatment for one to three weeks is required before clients realize a therapeutic effect from Buspar, and this is frequently too long to wait for individuals suffering acute anxiety.

Table 7.5 lists the trade and generic name, side effects, and other information for Buspar.

### Table 7.5 Azaspirone

| Trade Name | Generic Name | Side Effects | Other Facts |
|---|---|---|---|
| Buspar | Buspirone | Agitation, dizziness, drowsiness, headaches, G.I. upset, insomnia | Ineffective for both panic disorder and very severe anxiety. Adverse interaction with MAOIs. |

## Barbiturates

Barbiturates comprise the oldest class of antianxiety agents, but they have largely been replaced by the benzodiazepines and Buspar. Barbiturates are highly addictive, very sedating, and extremely dangerous because of their potential for lethal overdose as well as for severe, often life-threatening withdrawal symptoms associated with their discontinuation. All barbiturates cause dangerous interactions when ingested with alcohol. Elderly clients sometimes react paradoxically to barbiturates with confusion, excitement, or depression.

Table 7.6 on the following page lists trade and generic names, side effects, and other information for barbiturates.

**Table 7.6 Barbiturates**

| Trade Name | Generic Name | Side Effects | Other Facts |
|---|---|---|---|
| Butisol | Butabarbital | Agitation, anxiety, ataxia, confusion, G.I. upset, hallucinations, headaches, insomnia, nightmares, somnolence, swelling (of lips, cheeks, eyelids), thinking abnormalities | "Hangover" when drug wears off |
| Nembutal | Pentobarbital | Agitation, anxiety, ataxia, confusion, G.I. upset, hallucinations, headaches, insomnia, nightmares, somnolence, swelling (of lips, cheeks, eyelids), thinking abnormalities | "Hangover" when drug wears off |
| Pheno-barbital | Pheno-barbital | Agitation, anxiety, ataxia, confusion, G.I. upset, hallucinations, headaches, insomnia, nightmares, somnolence, swelling (of lips, cheeks, eyelids), thinking abnormalities | Hypersensitivity may produce severe, potentially fatal skin condition. Liver damage may occur. |

## Benzodiazepines

The benzodiazepines have less risk of lethal overdose and fewer side effects than the barbiturates and are therefore widely prescribed for anxiety. Approximately 10 percent of the U.S. population uses benzodiazepines each year (Kaplan & Sadock, 1996). Several of these medications have been developed specifically to treat insomnia and are known as hypnotics, hypnoanxiolytics, or sedative-hypnotics.

Discontinuation of benzodiazepines may be associated with severe and sometimes life-threatening withdrawal symptoms, including abdominal cramps, convulsions, fever, muscle aches, sweating, tremor, and vomiting (Maxmen & Ward, 1995b). Kaplan and Sadock (1996) write that the severity of withdrawal symptoms is positively correlated with six factors: potency of the particular drug; duration of use; dose level; abruptness of discontinuation; presence of panic disorder; and personality style having passive-dependent, histrionic, or somatizing traits. Abrupt withdrawal should be avoided, and doses should be tapered when these medications are discontinued.

Benzodiazepines are associated with several other problems. First, they produce a *rebound effect,* which is a temporary recurrence and worsening of pre-treatment symptoms (Kaplan & Sadock, 1996). The rebound effect is especially troublesome when the initial target symptom was insomnia, and the insomnia returns with increased intensity when the medication is discontinued. Second, class-specific side effects of benzodiazepines include excessive sedation, drowsiness, and difficulty concentrating, which can cause increased risk of injury from

### Table 7.7 Benzodiazepines

| Trade Name | Generic Name | Side Effects | Other Facts |
|---|---|---|---|
| Ativan | Lorazepam | Dizziness | |
| Dalmane | Flurazepam | Anxiety, ataxia, coma, dizziness, G.I. upset, skin rash | Hypnotic for treatment of insomnia; adverse interaction with alcohol |
| Halcion | Triazelam | Aggression, amnesia, ataxia, daytime anxiety, dizziness, drowsiness, G.I. upset, hallucinations uninhibited bizarre behavior | Hypnotic for treatment of insomnia |
| Klonopin | Clonazepan | Confusion, depression, hair loss, skin rash | Also for treatment of aggression, bipolar illness, and Tourette's |
| Librium | Chlordiazepoxide | Confusion, dizziness, EKG abnormalities, skin rash. | Slow onset |
| Restoril | Tenazepam | Confusion, dizziness, drowsiness, fatigue, G.I. upset | Hypnotic for treatment of insomnia |
| Serax | Oxazepam | Dizziness | Slow onset |
| Tranxene | Clorazepate | Dizziness, headache, skin rash | |
| Valium | Diazepam | Constipation, dizziness, drowsiness | Rapid action |
| Xanax | Alprazolam | Dizziness, drowsiness, dry mouth, irritability, memory problems, stuttering | Good for treatment of panic disorder |

falls, especially in elderly clients (Kaplan & Sadock, 1996). Third, benzodiazepines have been associated with memory loss that resolves with discontinuation of the drug. Severity of these side effects varies with differences in potency, half-life, and absorption time of each particular medication.

Finally, the benzodiazepines have a moderate potential for producing dependency, addiction, and tolerance. Therefore, the lowest possible therapeutic dose should be prescribed, treatment duration should not be prolonged, and clients with a history of problematic substance use should be monitored carefully.

Table 7.7 lists trade and generic names, side effects, and other information for the benzodiazepines.

### Antihistamines

Although their primary use is in the treatment of allergic reactions, antihistamines are occasionally prescribed to reduce symptoms of anxiety. Antihistamines are highly sedating, have few side effects, and do not promote addiction, dependency, or tolerance.

Table 7.8 on the following page lists trade and generic names, side effects, and other information for the antihistamines.

### Table 7.8 Antihistamines

| Trade Name | Generic Name | Side Effects | Other Facts |
|---|---|---|---|
| Atarax, Vistaril | Hydroxyzine | Drowsiness, dry mouth, sedation, tremor | Adverse interactions with alcohol, analgesics, barbiturates, and narcotics |
| Benadryl | Diphenhydramine | Dry mouth, drowsiness, G.I. upset, itching, sedation, skin rash, sweating, thickening of respiratory secretions, urinary retention, wheezing | Use with extreme caution in clients with asthma, peptic ulcers, prostate problems, urinary problems. |
| Phenergan | Promethazine | Blood pressure changes, blurred vision, dizziness, dry mouth, drowsiness, G.I. upset, sedation, skin rash | Use with extreme caution in clients with asthma. |

## Beta-Blockers

Beta-blockers were developed to treat hypertension and cardiac arrhythmias, but they are sometimes prescribed to reduce the symptoms of anxiety. They effectively alleviate physical symptoms, such as rapid heartbeat or sweating, but are not quite as effective in reducing psychological symptoms, such as a feeling of impending doom or difficulty concentrating (Maxmen & Ward, 1995b). In addition, they can sometimes precipitate a depressive episode. Beta-blockers have been used successfully to treat performance anxiety (Yudofsky, Hales, & Ferguson, 1991).

Table 7.9 lists trade and generic names, side effects, and other information for the beta-blockers.

### Table 7.9 Beta-Blockers

| Trade Name | Generic Name | Side Effects | Other Facts |
|---|---|---|---|
| Catapres* | Clonidine | Depression, fatigue, sedation | Primary use is treatment for hypertension. Abrupt withdrawal may cause agitation, anxiety, headache, hypertension. |
| Inderal | Propanolol | Depression, fatigue, hypotension, sedation, tachycardia | Avoid in clients with asthma or diabetes. |
| Tenormin | Atenolol | Depression, dry eyes, headache, sexual dysfunction, sedation, skin rash, sore throat | Use with caution in clients with kidney disease. Abrupt withdrawal may cause cardiac problems. |

*Sometimes categorized as alpha agonist.

# MOOD STABILIZERS
# (ANTICONVULSANTS AND LITHIUM)

Mood stabilizers were the first class of psychotropic medications used both to treat symptoms of mental illness and to prevent its recurrence (Stahl, 1996). Lithium and anticonvulsants are used, either separately or combined, to alleviate the symptoms of mania and hypomania and to stabilize the abnormally high and low mood fluctuations characteristic of bipolar disorder (Maxmen & Ward, 1995b). Symptoms of mania and the manic phase of bipolar disorder may include decreased need for sleep, psychomotor agitation, racing thoughts, pressured speech, grandiosity, involvement in potentially self-destructive activities (such as indiscriminate shopping sprees), and impaired thinking.

In addition to bipolar illness, mania and unstable mood may be induced by medications, such as steroids and antidepressants, and by certain medical conditions, such as Creutzfeldt-Jakob disease, epilepsy, liver failure, multiple sclerosis, stroke, and thyroid disease (Morrison, 1997; Yudofsky, Hales, & Ferguson, 1991). Other indications for mood stabilizers include treatment of aggression, alcohol withdrawal, atypical psychosis, depression with psychotic features, hallucinations associated with chronic substance abuse, mania, panic attacks, schizoaffective disorders, schizophrenia, and seizures (Maxmen & Ward, 1995b).

Mood stabilizers are associated with potentially serious conditions, so clients taking these medications must be regularly monitored through blood tests. Levels of lithium that exceed the narrow therapeutic margin in the blood can be toxic and possibly lethal. Furthermore, because lithium is excreted directly through the kidneys, clients must be careful to maintain a steady water volume in the body by drinking plenty of fluids and avoiding exercise-induced dehydration. Anticonvulsants may be associated with serious or lethal blood and liver disorders, and clients must be vigilant about early warning signs, such as easy bruising, bleeding, fever, sores in the mouth, and abnormal blood count.

Table 7.10 on the following page lists the trade and generic names, side effects, and other information for the mood stabilizers.

# ANTIPSYCHOTICS

Antipsychotics are used to treat psychosis, a condition in which the client is unable to differentiate what is real from what is not real. Symptoms of psychosis may include delusions, hallucinations, or any other severely impaired reality testing, as well as behavioral manifestations such as aggression, agitation, belligerence, catatonia, or disorganized speech. Psychosis may be due either to mental disorders, such as schizophrenia, severe depression, bipolar illness, and substance abuse, or to physical problems, such as brain tumor, epilepsy, head injury, Huntington's disease, infection, parathyroid disease, kidney failure, stroke, syphilis, temporal lobe epilepsy, and Wilson's disease (Morrison, 1997). Symptoms of psychosis can also be induced by prescription medications, toxic chemicals, and other materials.

## Table 7.10  Mood Stabilizers

| Trade Name | Generic Name | Side Effects | Other Facts |
|---|---|---|---|
| Depakote | Valproic acid | Appetite decrease, G.I. upset, itching, menstrual changes, tardive dyskinesia | Potentially fatal liver damage may occur. Local irritation of mouth may occur if tablets are chewed before swallowing. |
| Klonopin | Clonazepam | Ataxia, depression, dizziness, drowsiness, dry mouth, fatigue, sexual dysfunction | Good treatment for anxiety, especially panic. Drug dependency and tolerance may develop. Severe withdrawal symptoms may occur with abrupt withdrawal. Adverse interactions with alcohol, analgesics, anti-anxiety agents, barbiturates, MAOIs. |
| Eskalith, Cibolith, Lithobid | Lithium | Acne, ataxia, confusion, drowsiness, dry mouth, G.I. upset, hair loss, hallucinations, memory problems, thyroid problems, tremor, urinary problems, weight gain | Narrow margin of safety; toxic or lethal if too high; monitor blood levels; regulate liquid intake and output. |
| Lamictal | Lamatrogine | Accidental injury due to incoordination, ataxia, blurred vision, cough, dizziness, double vision, flu-like symptoms, nausea, pharyngitis, rash, rhinitis, somnolence, vomiting | Use with caution in patients with renal impairment. Caution patients not to drive while taking Lamictal. Notify physician at first sign of rash. |
| Neurontin | Gabapentin | Ataxia, dizziness, double vision, fatigue, rhinitis, somnolence | Take at bedtime to decrease side effects. No known drug interactions. Discontinue gradually over one week. |
| Tegretol | Carbamazepine | Blurred vision, constipation, dizziness, dry mouth, drowsiness, unsteadiness | May precipitate mania. Severe and potentially fatal skin conditions may occur. Rare but serious blood condition (aplastic anemia) may occur; monitor with blood tests. Adverse interactions with MAOIs. |

Antipsychotics have been used successfully to reduce anxiety and to alleviate the positive symptoms of psychosis, such as agitation, delusions, hallucinations, disorganized thinking and speech, and bizarre behavior. However, until recently, these medications were ineffective in treating the negative symptoms, such as attention impairment, blunted affect, difficulty in abstract thinking, emotional and social withdrawal, lack of goal-directed behavior, limited motivation, passivity, and poverty of speech (Stahl, 1996; Wolski, 1995). In the past few years, several atypical antipsychotics have been developed that effectively target not only the positive symptoms but also the negative symptoms, thus enhancing interpersonal relationships and reducing the sense of isolation suffered by clients with psychotic disorders. These include Clozaril (clozapine), Risperdal (respiridone), Seroquel (quetiapine), and Zyprexa (olanzapine).

Other indications for antipsychotics include autism, debilitating nausea, delusional disorder, delirium, dementia, schizotypal personality disorder, and Tourette's syndrome. Antipsychotics are also highly effective when combined with antidepressants to treat mood disorders with psychotic features (Kaplan & Sadock, 1996; Maxmen & Ward, 1995a, 1995b; Pollack, Otto, & Rosenbaum, 1996; Reid, Balis, & Sutton, 1997).

Antipsychotics are associated with a constellation of adverse side effects, including blood pressure changes, cardiac problems, breast enlargement in both males and females, skin eruptions, and weight gain. Other class-specific side effects that are particularly problematic include muscle spasms (*acute dystonic reaction*), motor restlessness (*akathisia*), pseudo-Parkinsonism (*akinesia*), involuntary movements (*tardive dyskinesia*), and a life-threatening condition known as *malignant neuroleptic syndrome*. These side effects are discussed here:

1. Acute dystonic reaction involves acute cramping, rigidity, and twisting of a muscle group, often in the neck, face, or along the spine (Yudofsky, Hales, & Ferguson, 1991). Muscle spasms may pull the head to one side or arch the neck backward, causing grimacing, protrusion of the tongue, and immobility of the eyes (Wolski, 1995). Acute dystonic reaction occurs in 10 percent of individuals taking antipsychotic medications, usually within a week of beginning treatment, and responds well to medical treatment.

2. Akathisia refers to a feeling of motor restlessness, relieved only by pacing, swinging the legs, or jumping up and down, and is a common side effect of antipsychotics. It occurs in 40 percent of clients after a single dose of Haldol (haloperidol) and in as many as 75 percent of clients who take Haldol for one week (Gitlin, 1990). Akathisia is difficult to treat but may be alleviated by reducing the dose of antipsychotic medication (Wolski, 1995).

3. Akinesia is a condition that occurs weeks or months after initiation of treatment with antipsychotics; it mimics the symptoms of Parkinson's disease. It involves a masklike expression, shuffling gate, drooling, hand tremor, stiff joints, and slow movement (Gitlin, 1990). Akinesia responds well to treatment with anticholinergic medications (Wolski, 1995).

4. Tardive dyskinesia is an irreversible syndrome associated with all antipsychotic medications except Clozaril (clozapine). It involves involuntary, repetitive muscular movements, such as grimacing, sticking out the tongue, smacking the lips, blinking, turning the neck and head, and twisting the extremities. Occasionally the movements may become severe enough to obstruct breathing (Yudofsky, Hales, & Ferguson; Wolki, 1995). Tardive dyskinesia is permanent and untreatable and is correlated with dose size and duration of treatment. Therefore, psychiatrists strive to prescribe the lowest effective dose of antipsychotics, for the shortest time, to prevent this syndrome.

5. Malignant neuroleptic syndrome is a rare but potentially life-threatening reaction characterized by coma or confusion, blood pressure fluctuations, kidney failure, very high fever, muscle rigidity, rapid heartbeat, and respiratory distress. This condition requires emergency medical treatment because the mortality rate is about 20 percent (Wolski, 1995). Malignant neuroleptic syndrome is most likely to occur upon initiation of treatment with antipsychotics, following rapid increases in dose, or after prolonged treatment at a high dose.

Table 7.11 on the following pages lists the trade and generic names, side effects, and other information for the antipsychotics.

## PSYCHOSTIMULANTS

Psychostimulants are used to treat attention-deficit hyperactivity disorder (ADHD) in children by decreasing motor activity, impulsiveness, forgetfulness, emotional lability, concentration problems, and difficulty staying on task (Maxmen & Ward, 1995b). Maxmen and Ward report that 85 to 90 percent of children with ADHD respond favorably to at least one of the psychostimulants if two different medications are tried. When only one medication is tried, 70 to 80 percent of children experience some improvement. However, effective treatment of a child with ADHD requires multiple, integrated, interlocking interventions in addition to the use of psychotropic medication, including individual and family psychotherapy, social skills training, parent guidance, school intervention, and monitoring of the child's physical health by a pediatrician or family physician.

Class-specific side effects of psychostimulants include insomnia, appetite suppression, and growth suppression in children, although no research evidence indicates that growth suppression is permanent (Kaplan & Sadock, 1996). Exacerbation or precipitation of tics and Tourette's syndrome may occur with the use of psychostimulants.

Psychostimulants are also used to reduce symptoms of inattention, difficulty concentrating, impulsivity, and irritability in adults who have ADD (Maxmen & Ward, 1995b). Other indications include treatment of apathy in the elderly; chronic fatigue syndrome; narcolepsy; and treatment-resistant

## Table 7.11 Antipsychotics

| Trade Name | Generic Name | Side Effects | Other Facts |
|---|---|---|---|
| Clozaril | Clozapine | Anxiety, chest pain, constipation, convulsions, depression, dizziness, drowsiness, dry mouth, excess salivation, G.I. upset, headache, shortness of breath, skin rash, tachycardia, weight gain | Use with caution in patients with cardiovascular, eye, kidney, liver, prostate, or urinary problems. Potentially fatal blood disease may occur. Not associated with tardive dyskinesia. |
| Haldol | Haloperidol | Agitation, akinesia, akathisia, anxiety, depression, dizziness, dry mouth, excess salivation, fatigue, insomnia, nasal congestion, sedation, rigidity, tardive dyskinesia, weight changes | May be used to treat Tourette's syndrome. Adverse interaction with alcohol. |
| Loxitane | Loxapine | Akathisia, blurred vision, dizziness, dry mouth, fatigue, hair loss, headache, hypotension, nasal congestion, pseudo-Parkinsonism, tachycardia, tardive dyskinesia, tremor, urinary retention | Use with extreme caution in patients with epilepsy. |
| Mellaril | Thioridazine | Blurred vision, cardiac arrhythmias, constipation, dizziness, drowsiness, dry mouth, hypotension, nasal congestion, sexual dysfunction, skin rash, swelling, tardive dyskinesia, urinary retention, weight gain | Adverse interactions with alcohol, anesthetics, barbiturates, narcotics. Avoid insecticides. |
| Navane | Thiothixene | Akathisia, dizziness, drowsiness, dry mouth, hypotension, nasal congestion, pseudo-Parkinsonism, skin rash, tardive dyskinesia, urinary retention, weight gain | Adverse interactions with alcohol, anesthetics, barbiturates, narcotics. |
| Prolixin | Fluphenazine | Agitation, blurred vision, drowsiness, dry mouth, excess salivation, G.I. upset, insomnia, sexual dysfunction, tardive dyskinesia, weight gain | Adverse interactions with alcohol, epinephrine, hypnotics. Use with caution in patients with liver disease. |

(continued on next page)

**Table 7.11 Antipsychotics *(continued)***

| Trade Name | Generic Name | Side Effects | Other Facts |
|---|---|---|---|
| Risperdal | Risperidone | Constipation, dizziness, drowsiness, tardive dyskinesia, weight gain | Controls both positive and negative symptoms. Avoid alcohol. |
| Seroquel | Quetiapine | Dizziness, headache, drowsiness, dry mouth, tardive dyskinesia | Controls both positive and negative symptoms. Avoid alcohol. |
| Stelazine | Trifluoperazine | Akathisia, appetite decrease, cardiac arrest, constipation, dizziness, drowsiness, hair loss, headache, muscle weakness, pseudo-Parkinsonism, skin rash, tardive dyskinesia, tremor | Adverse interactions with alcohol, anesthetics, antianxiety agents, barbiturates, narcotics. |
| Thorazine | Chlorpromazine | Akinesia, blurred vision, constipation, depression, dry mouth, drowsiness, EKG abnormalities, menstrual changes, muscle weakness, Parkinsonism, skin pigmentation, tachycardia, tardive dyskinesia, weight gain | Adverse interactions with alcohol, anesthetics, antianxiety agents, anticoagulants, anticonvulsants, barbiturates, epinephrine, narcotics, lithium. |
| Trilafon | Perphenazine | Akathisia, blurred vision, constipation, dizziness, dry mouth, nasal congestion, Parkinsonism, somnolence, tardive dyskinesia, tremor, weight gain | Use with caution in patients with depression, kidney, or liver problems. Adverse interactions with alcohol, antihistamines, analgesics, barbiturates, opiates. |
| Zyprexa | Olanzapine | Akathisia, constipation, dizziness, personality disorder, postural hypotension, somnolence, tardive dyskinesia, weight gain | Controls both positive and negative symptoms. Avoid alcohol. |

depression associated with Alzheimer's disease, dementia, and AIDS (Kaplan & Sadock, 1996; Maxmen & Ward, 1995a; Pollack, Otto, & Rosenbaum, 1996; Reid, Balis, & Sutton, 1997; Stahl, 1996).

Morrison (1997) noted that abnormalities in ability to focus attention may be associated with physical conditions such as fibromyalgia, postconcussion syndrome, and delirium induced by a variety of medical illnesses.

Table 7.12 on the following page lists the trade and generic names, side effects, and other information for the psychostimulants.

**Table 7.12 Psychostimulants**

| Trade Name | Generic Name | Side Effects | Other Facts |
|---|---|---|---|
| Adderall | Dextroamphetamine | Appetite suppression, dizziness, growth suppression, insomnia, irritability, weight loss | Associated with psychological dependence and drug tolerance |
| Cylert | Pemoline | Appetite suppression, growth suppression, insomnia | Potential for causing serious liver damage |
| Ritalin | Methylphenidate | Appetite suppression, depression, dizziness, growth suppression, G.I. upset, insomnia, weight loss | Associated with psychological dependence and drug tolerance |

# CONCLUSIONS

Psychotropic medications alleviate a wide spectrum of psychological symptoms and are an integral component of the treatment of many mental and emotional disorders. However, they must be prescribed and monitored with great care, and the potential risks must be judiciously weighed against the potential benefits for every individual client.

Furthermore, while these medications "exert profound and beneficial effects on cognition, mood, and behavior, they often do not change the underlying disease process, which is frequently highly sensitive to intrapsychic, interpersonal, and psychosocial stressors" (Schatzberg & Cole, 1991, p. 2). Optimal treatment entails a holistic approach, with attention not only to the biological underpinnings of mental and emotional disorders but to all aspects of the person. As is true in all counseling situations, counselors and counselor interns caring for clients who are concurrently taking psychotropic medications should strive to use a comprehensive, biopsychosocial vantage point to address all realms of the client's life experience.

# REFERENCES

AKISKAL, H. (1989). The classification mental disorders. In H. I. Kaplan & B. J. Sadock (Eds.). *Comprehensive textbook of psychiatry* (Vol. 1, 5th ed.) (pp. 583–598). Baltimore: Williams & Wilkins.

BEITMAN, B. D. (1991). Medications during psychotherapy: Case studies of the reciprocal relationship between psychotherapy process and medication use. In B. D. Beitman & G. L. Klerman (Eds.), *Integrating pharmacotherapy and psychotherapy* (pp. 21–43). Washington, DC: American Psychiatric Press.

BRADLEY, S. S. (1990). Nonphysician psychotherapist–physician pharmacotherapist: A new model for concurrent treatment. *Psychiatric Clinics of North America, 13*(2), 307–322.

BUSCH, F. N., & GOULD, E. (1993). Treatment by a psychotherapist and a psychopharmacologist: Transference and countertransference issues. *Hospital and Community Psychiatry, 44*(8), 772–774.

CHILES, J. A., CARLIN, A. C., BEN-JAMIN, G. A., & BEITMAN, B. D. (1991). A physician, a nonmedical psychotherapist, and a client: The pharmacotherapy–psychotherapy triangle. In B. D. Beitman & G. L. Klerman (Eds.), *Integrating pharmacotherapy and psychotherapy* (pp. 105–117). Washington, DC: American Psychiatric Press.

GABBARD, G. O. (1990). *Psychodynamic psychiatry in clinical practice.* Washington, DC: American Psychiatric Press.

GITLIN, M. J. (1990). *The psychotherapist's guide to psychopharmacology.* Toronto: The Free Press, Macmillan.

KAHN, D. (1990). The dichotomy of drugs and psychotherapy. *Psychiatric Clinics of North America, 13*(2), 197–208.

KAPLAN, H. I., & SADOCK, B. J. (1996). *Pocket handbook of clinical psychiatry.* Baltimore: Williams and Wilkins.

KLERMAN, G. L. (1991). Ideological conflicts in integrating pharmacotherapy and psychotherapy. In B. D. Beitman & G. L. Klerman (Eds.), *Integrating pharmacotherapy and psychotherapy* (pp. 3–19). Washington, DC: American Psychiatric Press.

MARTINDALE, P. (1990). Perspectives on psychotherapy and pharmacotherapy in the public hospital. *Psychiatric Clinics of North America, 13*(2), 333–340.

MAXMEN, J. S., & WARD, N. G. (1995a). *Essential psychopathology and its treatment* (2nd ed., revised for DSM-IV). New York: W. W. Norton.

MAXMEN, J. S., & WARD, N. G., (1995b). *Psychotropic drugs: Fast facts.* New York: W. W. Norton.

MEDICAL ECONOMICS COMPANY. (1998). *Physicians' desk reference.* Montvale, NJ.

MORRISON, J. (1997). *When psychological problems mask medical disorders.* New York: Guilford.

PAYKEL, E. S. (1995). Psychotherapy, medication combinations, and compliance. *Journal of Clinical Psychiatry, 56* (Suppl. 1), 24–30.

PILETTE, W. L. (1988). The rise of three-party treatment relationships. *Psychotherapy 25*(3), 420–423.

POLLACK, M. H., OTTO, M. W., & ROSENBAUM, J. F. (1996). *Challenges in clinical practice.* New York: Guilford.

REID, W. H., BALIS, G. U., & SUTTON, B. J. (1997). *The treatment of psychiatric disorders* (3rd ed., revised for DSM-IV). Bristol, Pennsylvania: Brunner/Mazel.

SCHATZBERG, A., & COLE, J. (1991). *Manual of clinical pharmacology* (2nd ed.). Washington, DC: American Psychiatric Press.

STAHL, S. M. (1996). *Essential psychopharmacology: Neuroscientific basis and practical applications.* Cambridge, England: Cambridge University Press.

WARD, N. G. (1991). Psychosocial approaches to pharmacotherapy. In B. Beitman & G. Klerman (Eds.), *Integrating pharmacotherapy and psychotherapy* (pp. 69–103). Washington, DC: American Psychiatric Press.

WOLSKI, R. L. (1995). Psychotropic medications. In S. Austrian, *Mental disorders, medications, and clinical social work* (pp. 240–259). New York: Columbia University Press.

WRIGHT, J. H., & THASE, M. E. (1992). Cognitive and biological therapies: A synthesis. *Psychiatric Annals, 22*(9), 451–458.

YUDOFSKY, S. C., HALES, R. E., & FERGUSON, T. (1991). *What you need to know about psychiatric drugs.* New York: Ballantine.

# 8

# Professional Challenges

During your internship you will most likely face a variety of challenging experiences that are inherent in the counseling profession. This chapter presents practical, basic guidelines for managing some of the anxiety-provoking situations and professional dilemmas that you may encounter for the first time as a counselor intern. We do not outline explicit treatment methods for specific clinical problems or disorders; rather, we address situations that frequently leave counselor interns feeling unsettled and wondering how to proceed. For information on ways to treat a particular clinical condition (for example, panic attacks), you should access other resources.

In fact, we strongly encourage you to use additional resources whenever you have questions concerning your interactions with clients. Discuss complex situations or dilemmas with your supervisor or other clinicians at your internship site; ask your counseling program professors for help; read professional books or journal articles that provide current clinical information; refer to the American Counseling Association's Code of Ethics and Standards of Practice (see Appendix E); and review policies and procedures outlined by your agency.

The particular needs, problems, and goals of your client, as well as the demands and constraints of the immediate clinical circumstances, will dictate your counseling style, treatment strategies and goals, and specific therapeutic interventions. The guiding principle in every case, however, involves working toward engaging the client in a strong therapeutic alliance. This therapeutic alliance enables you to collect data for assessing the client, to develop a compassionate and evolving understanding of the client, to formulate an appropriate treatment plan, and to begin helping (Shea, 1988). The following general goals provide a common platform for the engagement process and are well suited to most counseling situations:

- Provide a safe space, physically and psychologically.
- Establish rapport.
- Employ a collaborative stance.
- Understand your client's concerns contextually.
- Instill hope.
- Identify and mobilize your client's internal and external resources and support systems.
- Identify and mobilize your own resources and support systems.

The following sections list common clinical challenges in alphabetical order and provide recommendations for managing some of the perplexing, complicated, or stressful treatment situations you may experience during your counseling internship. We have also provided suggested readings wherever possible so that you may explore certain areas in more depth.

## ABUSE: REPORTING IT

*What should I do if a client tells me that a child or elderly person is being abused? How should I report the abuse?*

Helping professionals, such as physicians, nurses, teachers, social workers, and counselors, are mandated in many states to report suspected abuse of a child, adolescent under 18 years old, or elder. Reporting abuse is a difficult task because the counselor is likely to be distressed at hearing of the abuse, and clients often are upset by the need to report it. The abuser may be a family member whom the client loves and does not want to "get in trouble." The other family members may be upset and angry not only at the abuser but also at the client and the counselor. Many counselor interns describe their experience in making an abuse report as traumatic. If you find, during your internship, that you must report an incident of abuse, use your site supervisor, your counseling program supervisor, and other professors from your counseling department for support. Sharing your feelings with other students in your internship discussion group may also prove helpful.

As soon as you begin your internship placement, familiarize yourself with your state laws that pertain to reporting abuse, and collect the telephone numbers of the agencies that must be notified. Call the agency and ask what usually happens when they receive a report of abuse so that you will be able to share this information with the client, should the situation occur.

Consult with your site and counseling program supervisors to discuss what protocol to follow in case you suspect or learn that abuse is occurring. Your agency may have specific guidelines concerning the reporting of child abuse. You should follow all agency, county, and state procedures carefully, and document everything in detail.

When abuse is reported, agencies usually ask for the following information:

- The name, age, and address of the individual being abused
- In cases of child abuse, the name and address of the child's parent or guardian
- The name, age, and address of the abuser (if known)
- The exact nature of the abuse
- The location and date of the abuse
- Your name, position, and relationship to the abused individual or to the informant

Always make certain, during your initial discussion with clients concerning the limits of confidentiality, to discuss the fact that you are required by state law (if this is the case) to report all cases of suspected abuse. Then, if clients disclose abuse during their sessions, you will be able to remind them that counselors are legally mandated to report such situations, so that clients will not feel betrayed by a sudden breach of confidentiality. If your client asks you what will happen after the abuse is reported, present the facts gently and sensitively, but be honest about all procedures. Often a social worker will interview a child at school or will visit the home. Sometimes the individual who was abused will need a physical examination. The perpetrator may be charged. A child or elder may be removed from the home.

We have found it most helpful if you can encourage clients to make the telephone call to report abuse themselves while you sit nearby and offer support. If they are unwilling or unable to do this, it is a good idea for you to make the call yourself immediately, in the client's presence. In the case of a child, you should have the parent present if possible. Reporting abuse in the client's presence models being honest, taking action to keep people safe, and behaving in accordance with the law. Document in detail everything said during the phone call, as well as the fact that the client was present. We have found that these practices prevent later misrepresentations or accusations about what was reported. One of us, early in the counseling career, once called to report child abuse without the parent being present. The social worker who subsequently visited the home misstated the facts that we had reported, causing unnecessary rupture of our counseling relationship with the client.

## Suggested Readings on Abuse

BARNETT, O., MILLER-PERRIN, C., & PERRIN, R. (1997). *Family violence across the lifespan*. Thousand Oaks, CA: Sage.

BRERE, J. (1992). *Child abuse trauma*. London: Sage.

BRIERE, J., BERLINER, L., BULKLEY, J., JENNY, C., & REID, T. (Eds.). (1996). *The APSAC handbook on child maltreatment*. Thousand Oaks, CA: Sage.

BROWN, S. (1991). *Counseling victims of*

*violence.* Alexandria, VA: American Counseling Association.

GELLES, R. (1997). *Intimate violence in families.* Thousand Oaks, CA: Sage.

HERMAN, J. (1992). *Trauma and recovery.* New York: Basic Books.

KALICHMAN, S. (1993). *Mandated reporting of suspected child abuse: Ethics, law, and policy.* Washington, DC: American Psychological Association.

LEVINE, M., DOUCEK, H., & ASSOCIATES. (1995). *The impact of mandated*

*reporting on the therapeutic process.* Thousand Oaks, CA: Sage.

MALCHIODI, C. (1990). *Breaking the silence: Art therapy with children from violent homes.* New York: Brunner/Mazel.

WOLFE, D., McMAHON, R., & PETERS, R. (Eds.). (1997). *Child abuse: New directions in prevention and treatment across the lifespan.* Thousand Oaks, CA: Sage.

## ADOLESCENTS

*I will be seeing some adolescent clients during my internship, and I am wondering how to build a therapeutic alliance with them. What approach would be most helpful when I counsel adolescents?*

Adolescence, according to Erik Erikson (1963), is characterized by a search for identity and the consolidation of disparate elements of the self into an integrated whole that is congruent with self-perceptions and evaluation by others. Adolescents face some difficult developmental tasks as they struggle to find their way from childhood to adulthood. Corey and Corey (1992) described adolescence as a turbulent and possibly lonely period, characterized by many conflicts and paradoxes. During adolescence, boys and girls usually:

- Relinquish their dependency on the adults in their lives.
- Begin to separate from their families.
- Begin to make decisions that will affect their futures.
- Begin to relate to the opposite sex in new ways.
- Learn to interact with their peers successfully so that they will belong to a group.
- Must contend with powerful media and peer-group messages urging them to experiment with alcohol, drugs, and sex.
- Undergo the rapid physical development and emotional changes of puberty.

In addition, adolescents face the pressures and stressors of today's fast-paced, increasingly complex, and decreasingly family-oriented society. The result is often teenage depression, eating disorders, alcohol and drug abuse, sexual promiscuity and pregnancy, and juvenile delinquency (Dacey & Travers, 1991; Turner & Helms, 1991). The suicide rate for teenagers has increased over 300 percent since 1960; 13 adolescents kill themselves each day, and teen suicide attempts occur about 300,000 times each year, with guns or poison being the most common means used (Turner & Helms, 1991).

When you counsel adolescent clients, you will need to work hard to establish enough trust to build a therapeutic relationship. You may have more success if you are honest, reliable, and consistent and if you demonstrate your caring, support, interest, and respect.

Group therapy is often the most effective treatment modality for adolescents because of their strong allegiance to their peers (Corey & Corey, 1992; Dulcan & Popper, 1991). The therapy group provides a forum for teenagers to find that they are not alone with their problems, to experience a sense of belonging, to experiment with interpersonal interactions with their peers and the adult group leader(s), to learn to accept and provide honest feedback, and to explore values.

## Suggested Readings on Counseling Adolescents

COREY, G., & COREY, M. (1999). *Groups: Process and practice* (5th ed.). Pacific Grove, CA: Brooks/Cole.

GIL, E. (1996). *Treating abused adolescents.* New York: Guilford.

GULOTTA, T., ADAMS, G., & MONTEMAYOR, R. (1993). *Adolescent sexuality.* Newbury Park, CA: Sage.

KENDALL, P. (Ed.). (1991). *Child and adolescent therapy.* New York: Guilford.

KESSLER, J. (1996). *Psychopathology of childhood.* Englewood Cliffs, NJ: Prentice-Hall.

MORGAN, R. (1990). *Skills for living: Group counseling activities for young adolescents.* Alexandria, VA: American Counseling Association.

MUFSON, L., MOREAU, D., WEISSMAN, M., & KLERMAN, G.

(1993). *Interpersonal psychotherapy for depressed adolescents.* New York: Guilford.

O'NEILL, R., HORNER, R., ALBIN, R., SPRAGUE, J., STOREY, K., & NEWTON, J. (1997). *Functional assessment and program development for problem behavior.* Newbury Park, CA: Brooks/Cole.

SALO, M., & SHUMATE, S. (1993). *Counseling minor clients* (Vol. 4). Alexandria, VA: American Counseling Association.

VERNON, A. (1993). *Developmental assessment with children and adolescents.* Alexandria, VA: American Counseling Association.

# CHILDREN

*One of my clients is an eight-year-old child who appears anxious and does not say more than a few words during our sessions. How can I help this youngster?*

Counseling children often requires that you modify your professional role, your style, and your treatment modalities. You may need to be cognizant of the impact on your client of various systems, such as the family, the school, the neighborhood group of children, and sometimes the court or other social service agencies. Children are dependent on the adults around them, and therefore you may find that you need to act as an advocate for your client and as a liaison to all these powerful outside forces.

Most of the time the child client is brought to counseling or therapy because some adult in his or her life believes there is a problem, not because the child wants to be there. Therefore, you need to build a therapeutic relationship with the child's primary caregiver as well as with the child. In this respect, counseling children is more complex than counseling adults because you are faced with a triad rather than a dyad in your treatment relationship.

If you work with children during your internship, you need to be very familiar with developmental theory so that you can assess your client's cognitive, emotional, social, behavioral, and physical development. A thorough understanding of where the child stands developmentally allows you to:

- Relate to the child most effectively to establish trust and a therapeutic alliance.
- Determine whether problems or symptoms represent significant deviations from normal development.
- Identify goals for change.
- Formulate treatment plans in specific areas to help children "get back on track" developmentally.

Children often express themselves more easily through art and play therapy than through the more traditional "talking therapy." Even very verbal, bright children may not possess the vocabulary or linguistic ability to describe their perceptions, affective experiences, attitudes, or difficulties (Hughes & Baker, 1990). The child can more easily "play out" or draw powerful feelings, upsetting thoughts, worrisome situations, or past events. Play or art are helpful treatment modalities because they:

- Engage the child immediately by providing an enjoyable, nonthreatening activity and attractive materials.
- Afford the child emotional release (catharsis).
- Separate painful or troubling images from the child's self so they are not so secret any more.
- Increase the bond between yourself and the child as you share these feelings and experiences.
- Empower the child by giving him or her some symbolic control over situations where, in reality, he or she has little or no control.

For play therapy we recommend simple toys, such as a family of dolls, stuffed animals, hand puppets, blocks, cars and trucks, and perhaps a tea set, doctor set, or basic board game. Art materials should also be simple and geared to the child's developmental level, as well as to your own tolerance for messiness. You may select crayons, markers, play-dough, collage materials, or various types of paint. Some art materials are easier for the child to control than others, such as crayons as compared to watercolor paint, and this too is an important consideration.

To understand your child client's play and art, you will need to remember to think metaphorically (Dulcan & Popper, 1991). The play or art is a symbolic representation of the child's inner self and his or her perceptions of the world. We recommend that you keep your interpretations and interventions within the metaphor that the child has created. Do not connect your interventions too closely to the child because the child will feel threatened and communication will be hampered. For example, it is helpful to say "Tell me about the sad face you drew on this person" instead of "Do you feel sad like the person you drew in your picture?"

Recent studies (Lambert, Shapiro, & Bergin, cited in Rubin & Niemeier, 1993) indicate that cognitive behavioral approaches provide superior outcomes for behavioral problems in children. Cognitive behavioral techniques especially suited for counseling children include establishing small, attainable behavioral goals within a manageable time frame and providing positive reinforcement. For example, the child will remember to bring all homework assignments home from school for three consecutive days and will be rewarded with praise each day that the goal is accomplished, and with a small treat after three days of success. Goals can be increased and time periods can be gradually extended as the child achieves success. Parents and teachers need to be closely involved so that the child experiences consistent support and guidance in developing self-control.

Working successfully with a child client involves being creative and flexible in tailoring your therapeutic approach to your client's developmental level, as well as being sensitive to the child's unique personality and needs. Patience, warmth, a sense of humor, and an honest enjoyment of children are also helpful characteristics of the child therapist.

## Suggested Readings on Counseling Children

AXLINE, V. (1969). *Play therapy.* New York: Random House.

BRASWELL, L., & BLOOMQUIST, M. (1991). *Cognitive-behavioral therapy with ADHD children.* New York: Guilford.

BROMFIELD, R. (1992). *Playing for real: The world of a child therapist.* New York: Dutton.

DULCAN, M., & POPPEK, C. (1991). *Concise guide to child and adolescent psychiatry.* Washington, DC: American Psychiatric Press.

GIL, E. (1991). *The healing power of play.* New York: Guilford.

HUGHES, J., & BAKER, D. (1990). *The clinical child interview.* New York: Guilford.

KENDALL, P., & BRASWELL, L. (1993). *Cognitive-behavioral therapy for impulsive children.* New York: Guilford.

KLEPSCH, M., & LOGE, L. (1982). *Children draw and tell.* New York: Brunner/Mazel.

McWHIRTER, J., McWHIRTER, B., McWHIRTER, A., & McWHIRTER, E. (1998). *At-risk youth: A comprehensive response* (2nd ed.). Newbury Park, CA: Brooks/Cole.

ORTON, G. (1997). *Strategies for counseling with children and their parents.* Pacific Grove, CA: Brooks/Cole.

SCHAEFER, C., & O'CONNER, K. (Eds.). (1983). *Handbook of play therapy.* New York: Wiley.

THOMPSON, C., & RUDOLPH, L. (1997). *Counseling children* (4th ed.). Pacific Grove, CA: Brooks/Cole.

# CONFIDENTIALITY

*How can I explain the limits of confidentiality to my adult clients and still expect that they will trust me enough to disclose personal information? How can I explain confidentiality to children in a way they will be able to understand?*

Our litigious society, rather than our own personal values or professional ethical standards, often dictates the parameters of our counseling relationship (Gutheil & Gabbard, 1993). The sanctity of the counselor–client alliance and professional confidentiality are no longer absolute legal standards. Therefore, you must always disclose the limits of confidentiality to your clients at the outset of your counseling relationship. Find out your exact state regulations concerning the situations in which you will need to break confidentiality, and memorize these.

As a novice counselor, you may find the task of defining confidentiality to your clients awkward or threatening. However, most clients will be reassured to have this information because many of them will be wondering about the privacy of what they talk about during their counseling sessions. We recommend that you try writing out your explanation of confidentiality several different ways, and then practice saying it, by yourself, until you feel most comfortable. When talking to clients, we stress that strict confidentiality is maintained most of the time, and that confidentiality is of primary importance to us.

We explain confidentiality to our adult clients this way: "Everything we say to each other during our counseling sessions is absolutely private, except in three instances. By law, I am required to get help for you if you feel like hurting yourself, and I am required to report to the proper authorities if you tell me that you have plans to hurt another person or if I learn of any abuse of children or the elderly. In all of these situations, I would discuss my plans with you before talking to anybody else. My only motive would be to protect you and others from harm, to help keep people safe. I'd be glad to answer any questions or hear any comments you may have about what I've just said."

During your internship, and as long as you require supervision, you will also need to discuss your status as a counselor trainee. You may wish to say, "The state requires that all mental health professionals be supervised by more experienced helpers for a number of years to ensure the highest quality of care for all clients. I am currently being supervised by Dr. Smith, who is a Licensed Professional Clinical Counselor (or whatever title the supervisor holds). I will be reviewing your records with him from time to time so that he can offer suggestions concerning our counseling sessions. However, Dr. Smith will keep everything absolutely confidential."

For children, we modify our explanation of confidentiality a bit: "The things we talk about in here are mostly private, just between us, because it's important that you can tell me anything and everything you are feeling or thinking. But it's really important for me to keep you and other people safe, too. So if you tell me something that makes me worry that you might hurt yourself or somebody else, or if I hear that a child or older person is being hurt, then I would have to get some help. I would have to talk to some other adults who could help us handle the problem. But I would always talk it over with you first."

## ELDERS

*My internship site serves some elders. What can I do in counseling them that would be helpful?*

Approximately 15 percent of the elders in this country manifest at least moderate emotional problems, but only 2 percent receive mental health care (Turner & Helms, 1991). Three factors contribute to elders' low use of mental health services. First, there is a widely held, albeit mistaken, belief that depression and other distressing psychological states are a normal part of aging. Second, many elders have limited, fixed incomes and problems with transportation that reduce access to mental health care. Third, many mental health professionals are reluctant to treat elders due to misconceptions based on limited training (Lasoski, 1986) or personal discomfort with issues of growing old and dying (Butler, 1983).

In counseling elders, keep in mind that depression and anxiety are not inevitable conditions of aging. On the contrary, counseling and psychopharmacological intervention are highly effective in treating mental and emotional disorders of elders (Zivian, Larsen, Knox, Gekoski, & Hatchette, 1993).

Elders often must cope with loss in many areas. For example, they may face the following:

- Decline in physical strength and vigor, sensory acuity (vision, hearing, and taste), youthful appearance, and sexual potency (in males)

- Loss of power, status, and financial security associated with retirement from career role

- Decreased independence and autonomy due to health problems or limited funds

- Decreased social support system because friends, siblings, and spouse or significant other may be gone

- Discomfort of certain degenerative illnesses, such as arthritis, associated with aging

In addition, elders confront their own mortality and the fact that "time is running out." They reminisce about the past and try to reconsider experiences and conflicts in order to derive some sense of meaning and continuity in their lives. Robert Butler (Butler, 1983; Butler & Lewis, 1981) writes that this process of life review is an integral component of the life cycle.

During your internship, you can help your elder clients by assisting them in their life review. You can listen empathically to your clients' stories of the past and understand that this narration is not simply "rambling talk" or "resistance to discussing real issues," but rather a valuable and crucial aspect of counseling elders.

It is imperative to refer your elder clients to a physician experienced in gerontology as a prerequisite to counseling. Underlying organic disease causes many cases of depression or impaired thinking. A psychiatric consultation may be required, as well, because psychotropic medications are useful in alleviating many disturbing psychological symptoms (such as insomnia or paranoia) in elders.

## Suggested Readings on Counseling Elders

CAVANAUGH, J. (1997). *Adult development and aging* (3rd ed.). Newbury Park, CA: Brooks/Cole.

KNIGHT, B. (1996). *Psychotherapy with older adults* (2nd ed.). Thousand Oaks, CA: Sage.

LEIBLUM, S., & SEGRAVES, T. (1989). Sex therapy with aging adults. In S. Leiblum & R. Rosen (Eds.), *Principles and practice of sex therapy: Update for the 90's* (pp. 352–381). New York: Guilford.

MYERS, J. (1989). *Adult children and aging parents*. Alexandria, VA: American Counseling Association.

MYERS, W. (1991). *New techniques in psychotherapy of older patients*. Washington, DC: American Psychiatric Press.

SADAVOY, J., LAZARUS, L., & JARVIK, L. (Eds.). (1991). *Comprehensive review of geriatric psychiatry*. Washington, DC: American Psychiatric Press.

TICE, C., & PERKINS, K. (1997). *Mental health issues and aging: Building on the strengths of older people*. Newbury Park, CA: Brooks/Cole.

## GROUP VERSUS INDIVIDUAL THERAPY

*One of my clients is shy and has difficulty with interpersonal interactions. How can I decide whether group therapy would be more helpful for her than individual counseling?*

Individual therapy does not preclude your client's participation in a group; individual and group therapies are often effective concomitant treatment modalities. Yalom (1985) emphasizes that groups are suitable for almost all clients as long as the groups' focus, structure, and composition are carefully matched to the clients' needs.

Corey and Corey (1992) suggest that group therapy is an especially appropriate format for adolescent clients because peer interactions are so important during the teenage years. In addition, they write that groups offer an advantage because the adolescent task of separating from adults can sometimes interfere with the development of a positive therapeutic relationship in individual counseling. Group therapy may also be the preferred treatment modality in instances where individual therapy has been frequently disrupted by intensely negative transference (for example, in some clients with borderline personality disorder) because the presence of other people in the group somewhat dilutes the intensity of the counselor–client relationship (Yalom, 1985).

Certain general types of clients may not be suitable candidates for group therapy. Usually clients who have brain injuries, drug or alcohol addictions, paranoid ideation, florid psychotic disorders, or sociopathic behaviors should not be referred to groups (Yalom, 1985). Yalom writes that it is also wise to eliminate clients who are unlikely to attend the group regularly, either because they live too far from the meeting place or because they must travel frequently for business. Absences or "drop-outs" are disruptive and potentially damaging to the other members of the group.

During your internship, most of your clients will benefit from a combination of group and individual therapies. Facilitating groups is usually an enjoyable learning experience for counselor interns.

## Suggested Readings on Group Therapy

COREY, G. (1991). *Theory and practice of counseling and psychotherapy* (4th ed.). Pacific Grove, CA: Brooks/Cole.

COREY, G. (2000). *Theory and practice of group counseling* (5th ed.). Pacific Grove, CA: Brooks/Cole•Wadsworth.

COREY, G., & COREY, M. (1997). *Groups: Process and practice* (5th ed.). Pacific Grove, CA: Brooks/Cole.

COREY, G., COREY, M., CALLAHAN, P., & RUSSELL, M. (1997). *Group*

*techniques* (2nd ed.). Newbury Park, CA: Brooks/Cole.

JACOBS, E., MASSON, R., & HARVILL, R. (1998). *Group counseling strategies and skills* (3rd ed.). Newbury Park, CA: Brooks/Cole.

YALOM, I. (1985). *Theory and practice of group psychotherapy*. New York: Basic Books.

## HOMICIDE: ASSESSING RISK

*One of my clients made a statement about wanting to kill someone. What should I do?*

Homicidal ideation always needs to be explored immediately and thoroughly because there are legal as well as moral implications when clients talk about killing anyone. Even children who express homicidal feelings must be taken seriously; there are many cases of children committing murder. Although there is no sure way to predict homicide, the most significant risk factors include "a history of violence, poor impulse control, or attempted homicide" (Shea, 1988, p. 437).

If your client verbalizes homicidal ideation during your internship, you should immediately consult with your site supervisor or another on-site senior clinician. As a counselor intern, you do not have enough clinical experience to assess the lethality of homicidal ideation. In case you still feel too uncomfortable to continue working with the client even after the clinician has decided there is minimal danger of anyone being hurt, discuss the situation and explore alternative treatment options with your supervisor. Make certain to document all interactions and consultations and to let your counseling program supervisor know what is going on.

## NONCOMPLIANCE WITH MANDATED COUNSELING

*What should I do about a client who is noncompliant with mandated treatment? One of my clients, a 19-year-old male, was required by his college adviser to attend counseling sessions for three months. He came only one time and has not been back.*

Some clients, especially those who are mandated to seek counseling by the court or other institutions, will be noncompliant. You should discuss the situation with your supervisor and adhere to agency guidelines. We recommend that you attempt to contact the client to encourage him or her to return to counseling. However, if you are unsuccessful in reaching the client, or if he or she refuses to return, then you should send a letter to the authorities that initially mandated counseling, explaining the problem and your efforts to resolve it. Both you and your supervisor will need to sign this letter. Keep a copy of the letter at your internship site, in the client's chart, and keep another copy for yourself. Also, document everything in detail, including dates and times of telephone calls, in the client's chart.

## PSYCHOSIS

*I am a counselor intern in an agency that serves clients with serious psychiatric disorders. What should I do if my client is actively psychotic?*

If your client population includes individuals with schizophrenia, bipolar illness, or other serious psychiatric disorders, you should discuss with your site supervisor, as soon as you begin your placement, what to do if you encounter a client who is psychotic. Most often, such clients require adjustment of their psychotropic medications. They may have stopped taking their medicine because of uncomfortable side effects. Sometimes, however, you may see a client who comes in with a first episode of psychotic illness. All clients showing signs of active psychotic process should be referred to a psychiatrist for evaluation. They usually need to be hospitalized, and they will be placed on a medication regimen that will require about 10 days to 2 weeks until therapeutic blood levels have been reached.

When interacting with psychotic clients, you should be supportive, calm, and very concrete. Try not to ask too many questions because many of these clients are suspicious and feel easily threatened and overwhelmed. Do not try to convince clients in an outpatient setting that their hallucinations exist only in their own minds. If you are in an inpatient setting, with staff approval you may be able to do gentle reality testing, offering reassurance. For example, you can say, "It must be frightening to hear those voices, but they really are part of your illness, and your medication is going to help with that."

Remember that clients who are suffering psychotic illness do not see, hear, interpret, or understand the world in the same way that you do. They are out of touch with reality and may behave very unpredictably or aggressively until their psychosis is under control or in remission. Your safety should be your first priority.

### Suggested Readings on Serious Psychiatric Disorders

BASCO, M., & RUSH, A. (1996). *Cognitive-behavioral therapy for bipolar disorder.* New York: Guilford.

GOSSETT, J., & GROB, M. (1991). *Psychiatric treatment.* Washington, DC: American Psychiatric Press.

GREDEN, J., & TANDON, R. (1991). *Negative schizophrenic symptoms.* Washington, DC: American Psychiatric Press.

HATFIELD, A. (1990). *Family education in mental illness.* New York: Guilford.

HATFIELD, A., & LEFLEY, H. (1994). *Surviving mental illness.* New York: Guilford.

KINGDON, D., & TURKINGTON, D. (1994). *Cognitive-behavioral therapy of schizophrenia.* New York: Guilford.

SEEMAN, M., & GREBEN, S. (1990). *Office treatment of schizophrenia.* Washington, DC: American Psychiatric Press.

SHEA, S. (1988). *Psychiatric interviewing: The art of understanding.* Philadelphia: Saunders.

TALBOT, J. (1987). *A family affair: Helping families cope with mental illness.* Washington, DC: American Psychiatric Press.

## REFERRALS

*How do I refer a client who will need to continue treatment after leaving my agency?*

In today's managed health care environment, the HMO or insurance company will usually restrict referrals to clinicians who are on the "provider panel." In these cases your client will choose a counselor from that list. If the client is not covered by a managed care organization, your agency will outline referral procedures and will usually have a list of community resources available. Your supervisor may be able to offer you suggestions concerning referrals and after-care planning for your client. You can also consult with other clinicians at your site, as well as your counseling program professors. Often human service agencies that do not provide the particular service needed by your client will offer names and telephone numbers of other organizations that do. We suggest that you begin assembling your own network of outside therapists and community resources as soon as you can. Collect names of professionals you meet at workshops or meetings; ask colleagues to share names of clinicians they know and respect; call colleges and universities in your area that offer graduate programs in counseling, social work, or psychology to ask for referral sources.

To begin the referral procedure, explore your client's needs and preferences in detail and try to accommodate your client as much as possible so that he or she will be more likely to follow through on the after-care plans. For example, consider whether your client requires help with emotional or interpersonal issues, family situations, career or work-related problems, or educational concerns. Perhaps your client needs a case manager who can coordinate services and help with such things as transportation or housing, or a psychiatrist, who can prescribe and monitor medications or manage a serious psychiatric disorder. Consider resources such as Alcoholics Anonymous or other community support groups. If your client has a specialized problem related to health or aging, you may call local hospitals, medical schools, or universities to inquire about making referrals to gerontological services, rehabilitation, or psychiatry departments.

The next step in your referral procedure involves choosing several resources that seem to be good possibilities. Share pertinent information with your client, but encourage your client to make the final selection. Explain to your client that he or she may want to try out more than one of these helpers to find the best fit. You may need to offer support for some clients and actually be there, standing by, as they telephone to set up their first appointments. Finally, complete necessary paperwork and obtain signed forms for release of information, so that you can send client records to the clinician or agency that will provide care.

## RESISTANCE

*My client has been arriving late for our sessions or sometimes missing her appointments. When she does come for counseling, she discusses everything except her own issues. She often will talk about her neighbors, the weather, or a television program in great detail. I know this client is being resistant, but I am not sure why or what to do about it.*

Corey (1991) defines resistance as "anything that works against the progress of therapy. Resistance refers to any idea, attitude, feeling, or action (conscious or unconscious) that fosters the status quo and gets in the way of change" (p. 122). Change is generally uncomfortable. Resistance comes into play when the client tries to avoid dealing with painful, threatening, or unpleasant thoughts and feelings. You can therefore understand resistance as a defense against anxiety and as a protective coping mechanism, albeit a maladaptive one in counseling. As a novice counselor, you will need to remember that your client is not simply being uncooperative and that the resistant behavior is not meant to frustrate you (Corey, 1991; Kahn, 1991). The resistance needs to be appreciated as a meaningful, unconscious, defensive dynamic that can help you better understand your client's inner world. Shea (1988) explains:

A seasoned clinician eventually recognizes resistance not as a demon but as an odd ally of sorts, offering an opportunity for insight. The presence of resistance serves to alert the clinician that the anxieties and defenses of the patient are near the surface. If attended to sensitively, resistance is a pathway to understanding (p. 500).

One strategy in addressing your client's resistance is encouraging your client to explore the source of the resistant behaviors. When your client becomes aware of the meaning and need for such behaviors as being late, skipping sessions, remaining silent for long periods, asking questions about your personal life, and making small talk, he or she will be more ready to acknowledge and master the underlying painful issues that he or she has been avoiding. Often, just your empathic comment may help your client overcome resistance to discussing difficult issues: "It is hard for you to get started talking about your father, isn't it?"

### Suggested Readings on Client Resistance

ARONSON, M., & SCHAKFMAN, M. (1992). *Psychotherapy: The analytic approach.* Northvale, NJ: Jason Aronson.

BENJAMIN, A. (1987). *The helping interview.* Boston: Houghton Mifflin.

GABBARD, G. (1990). *Psychodynamic psychiatry in clinical practice.* Washington, DC: American Psychiatric Press.

KAHN, M. (1991). *Between therapist and client.* New York: Ereeman.

McCOWN, W., & JOHNSON, J. (1992). *Therapy with treatment resistant families.* New York: Haworth.

SHEA, S. (1988). *Psychiatric interviewing: The art of understanding.* Philadelphia: Saunders.

TRAVERS, J., & STERN, M. (Eds.). (1985). *Psychotherapy and the uncommitted patient.* Binghamton, NY: Haworth.

URSANO, R., SONNENBERG, S., & LAZAR, S. (1991). *Concise guide to psychodynamic psychotherapy.* Washington, DC: American Psychiatric Press.

## SPECIAL SITUATIONS

*How can I be most helpful to clients who are coping with situations with which I have no experience or training?*

Having a physical or mental disability or a chronic medical illness can be associated with significant psychological stress. Furthermore, a wide range of medical conditions can cause depression, anxiety, or even psychosis (Morrison, 1997). Thus psychological disorders or emotional upset often occur concomitantly with physical and mental challenges and with chronic illnesses. You should relate to your clients who have or are coping with special situations (for example, problems related to HIV, spinal cord injury, cancer, chronic obstructive pulmonary disease, or mental retardation) in the same way you relate to all your clients. Demonstrate your empathy, interest, respect, nonjudgmental acceptance, genuineness, caring, trustworthiness, and desire to help. The establishment of a therapeutic alliance provides "a humanizing social attachment and reciprocal lines of communication" that are healing for all clients, regardless of their special situations (Bihm & Leonard, 1993, p. 231).

You will need to be creative and flexible and to ask your client what he or she finds most helpful. An honest approach is best. You can say, for example, "I haven't counseled anyone before who has multiple sclerosis, but I'm very glad to be working with you. I hope you will feel comfortable enough to tell me just how I can help. I am going to work hard to understand how you feel and how things are for you."

You should learn as much as possible about the unique needs and problems associated with your client's special situation. You can do research at the library or on the Internet and contact organizations that offer specialized information (for example, the Muscular Dystrophy Association, the American Cancer Association, or the National Institute of Health). We suggest also that you consult with your site and counseling program supervisors, who may be able to share their insights and experiences with you.

## Suggested Readings on Clients with Special Needs

BAUMEISTER, R. (1991). *Meanings of life.* New York: Guilford.

CANTWELL, D., & BAKER, L. (1991). *Psychiatric and developmental disorders in children with communication disorder.* Washington, DC: American Psychiatric Press.

COSTA, P., & VANDENBOOS, G. (1990). *Psychological aspects of serious illness.* Washington, DC: American Psychological Association.

GOODNIK, P., & KLIMAS, N. (1993). *Chronic fatigue and related immune deficiency syndromes.* Washington, DC: American Psychiatric Press.

HANSON, R., & GERBER, K. (1990). *Coping with chronic pain.* New York: Guilford.

MACKLIN, E. (1989). *AIDS and families.* New York: Haworth.

ROURKE, B., & FUERST, D. (1991). *Learning disabilities and psychosocial functioning.* New York: Guilford.

SCHOVER, L., & JENSON, S. (1988). *Sexuality and chronic illness.* New York: Guilford.

SOHLBERG, M., & MATEER, C. (1989). *Introduction to cognitive rehabilitation.* New York: Guilford.

STEWART, D., & STOTLAND, N. (1993). *Psychological aspects of women's health care.* Washington, DC: American Psychiatric Press.

## SUICIDE: ASSESSING RISK

*My client appears to be very depressed and has made the comment, "I really think I'd be better off dead." How do I decide what to do next?*

All suicidal statements need to be taken seriously and thoroughly evaluated. About two-thirds of the people who express suicidal feelings eventually do kill themselves, and completed suicide occurs in over 15 percent of persons suffering from depression (Dubovsky, 1988).

Some interns are afraid that if they ask a question such as "Have you been thinking about hurting or killing yourself?" they will be giving the client the idea. *Asking about suicidal thoughts will not encourage your client to commit suicide.* Most of the time clients are relieved to be asked because they have ambivalent feelings and confusing thoughts about suicide and want to be helped to feel better. As a counselor intern, you should attempt to elicit suicidal ideation with every client, and you should certainly explore and assess all suicidal thoughts or statements to determine lethality so that you can help ensure your client's safety.

The client who verbalizes suicidal wishes is expressing the intensity of his or her feelings of hopelessness, helplessness, loneliness, or other pain. You should respond with empathy and sensitivity to your client's distress while evaluating the following hierarchy of risk factors:

- Intent
- Plan
- Means

First, ask about the client's intent to kill himself or herself: "Have you been feeling like hurting or killing yourself?" It is also important to inquire about the intensity of suicidal thoughts or feelings: "How often do you have these thoughts of killing yourself?" or "How long did you stand on the bridge wondering whether to jump?"

Second, ask whether the client has a concrete plan: "Have you been thinking of how you would kill yourself? Tell me about what you would do."

Third, ask whether the client has the means to carry out this plan: "Do you have access to a gun?" or "How many of your pills have you saved up by now?" Clients who have suicidal ideation along with intent to kill themselves, a definite plan, and the means to execute this plan usually present a high risk.

We would also like to note several other situations that should alert you to suicidal risk in your client:

- The client has been taking antidepressants for several weeks to relieve symptoms of a mood disorder and has started to feel better. Often, paradoxically, when clients begin to feel better, they have enough energy to carry out their plans to commit suicide.
- The client has made previous suicidal gestures or attempts.
- The client has no social support system or *believes* that he or she has no social support system. Sometimes the client's perceptions regarding the availability of others are incorrect, but keep in mind that it is this belief that is acted upon.
- Family members, friends, classmates, or coworkers have committed suicide. Anniversary dates of those events may be especially difficult.
- The depressed client appears suddenly calm or speaks of feeling relieved, of having found a solution to his or her pain. Always ask about the source of relief.

- The client gives away possessions or makes arrangements concerning legal or financial affairs.

- The client is an adolescent who is very distraught over a breakup with a boyfriend or a girlfriend. Many adolescents are impulsive and have not yet developed either good judgment or the ability to foresee the long-term consequences of acting on their immediate feelings.

- The client has been suffering from a serious medical illness or chronic health problem.

- The client has a diagnosis of schizophrenia, alcohol dependency, or borderline personality disorder.

- Your intuition tells you that something is not right or that the client is upset enough to hurt himself or herself.

Once you have identified, explored, and assessed suicidal ideation and have determined that there is significant risk, what should you do next? First, be certain to let your supervisor know about the situation before you allow the client to leave the office. Depending on the circumstances and your experience, your supervisor may want to interview the client or to review your plans. Next, mobilize the client's own resources, such as religious values, spiritual beliefs, family, friends, and the healthy part of the self that is seeking help. Ask your client what has been helpful in resisting the impulse to kill himself or herself so far, and emphasize the importance of these factors. Explain that depression and suicidal feelings are time-limited and treatable and that the client will be able to get help to feel better.

If you and your supervisor ascertain that it is safe for the client to go home, secure a signed behavioral contract in which the client agrees that (1) he or she is able to stay safe and (2) he or she will telephone for help or go to the hospital emergency room if suicidal feelings become unmanageable. Provide a 24-hour telephone number for your client, in accordance with agency regulations, such as a crisis hot line, psychiatric emergency service, or clinician on call. Maintain a close, supportive relationship with the client and increase the frequency of your counseling sessions, even seeing the client daily if necessary, until the crisis subsides.

If the client cannot agree to keep herself or himself safe, or if your gut feeling tells you that the client is still in trouble, use other resources. Ask the client if family members or friends can be called to help. Consult with your supervisor or other clinicians on site. If hospitalization is indicated, follow agency procedures. Try to encourage voluntary hospitalization for the seriously suicidal client by discussing the need for "getting all the help we possibly can to keep you safe." Unfortunately, involuntary hospitalization is sometimes necessary. If this is the case, your site supervisor should be close by to assist. Follow agency guidelines, hospital policies, and county or state regulations carefully. In all instances, carefully document the client's mood, affect, behavior, and exact statements, as well as your interventions and plans.

## Suggested Readings on Suicidal Behavior

BERMAN, A., & JOBES, D. (1991). *Adolescent suicide: Assessment and intervention.* Washington, DC: American Psychological Association.

BONGAR, B. (1991). *The suicidal patient: Clinical and legal standards of care.* Washington, DC: American Psychological Association.

JACOBS, D. (Ed.). (1992). *Suicide and clinical practice.* Washington, DC: American Psychiatric Press.

LANN, I., MOSCICKI, E., & MARIS, R. (Eds.). (1989). *Strategies for studying suicide and suicidal behavior.* New York: Guilford.

LEENAGERS, A., MARIS, R., McINTOSH, J., & RICHMAN, J. (1992). *Suicide and the older adult.* New York: Guilford.

MCINTOSH, J., SANTOS, J., HUBBARD, R., & OVERHOLSER, J. (1994). *Elder suicide.* New York: Guiford.

WHITAKER, L., & SLIMAK, R. (Eds.). (1990). *College student suicide.* New York: Haworth.

# TRANSFERENCE AND COUNTERTRANSFERENCE

*I have been counseling a client for several weeks and I thought that we had a good therapeutic relationship. Recently, however, she has become angry with me during our sessions and accuses me of rejecting her. I actually like this client very much, but now I often want to avoid our sessions because I find them unpleasant, and I have begun distancing myself emotionally from my client. I am not sure what is going on or how to repair our relationship.*

You are experiencing psychological phenomena known as transference and countertransference. These concepts constitute central dynamics in psychoanalytic therapy; however, appreciating the importance of transference and countertransference is pertinent to your understanding of all interpersonal interactions, including all therapeutic relationships.

*Transference* refers to the transfer of feelings, attitudes, fears, wishes, desires, and perceptions that belonged to our past relationships onto our current relationships, so that the people in our present lives become the focus of these thoughts and emotions from long ago. Transference also refers to the tendency of the client to transfer feelings from other current relationships onto the counselor, thereby reexperiencing these feelings in the counseling session.

Transference occurs for two reasons. First, the human mind constantly compares incoming information with both conscious and unconscious memories in an effort to find some pattern match and to organize new information in a meaningful way (Eisengart & Faiver, 1996). Therefore, our perceptions of present situations are categorized according to, and linked to, our memories of past situations.

Second, all people seem to have a psychological need to repeat the past in an effort to master that which was difficult or emotionally painful. Because psychological development invariably involves difficulty and pain, this "compulsion to repeat," and the transferences that result, are ubiquitous human experiences (Ursano, Sonnenberg, & Lazar, 1991, p. 43).

In a counseling relationship, the transference can become especially intense and can feel very real to the client because counseling often evokes powerful feelings and conflicts from the past. In the question posed at the beginning of this section, your client is actually experiencing your behavior as rejecting, just as she experienced rejection from some other significant person in her life, perhaps from a parent. Your client is probably also reexperiencing the same hurt feelings, pain, and anger toward you as she did toward that other person, long ago. You can help her by demonstrating that, in reality, you do care for her, that you do want to "be there" for her, and that you are trying hard to understand her feelings. You may also explain the concept of transference and encourage her to explore where her feelings may be originating.

*Countertransference* is similar to transference and refers to the counselor's emotional responses to the client. Countertransference may be due to the counselor's unconscious feelings, thoughts, and conflicts from the past that are evoked by the present clinical situation. However, countertransference can also be interpreted as a response to the client's transference, and it can be used by the counselor to gain insight.

Ursano, Sonnenberg, and Lazar (1991) explain that countertransference can be experienced in two ways. In the first, known as *concordant countertransference,* the counselor understands and feels empathy for the client's emotions. In the second, known as *complementary countertransference,* the counselor's feelings toward the client are similar to those of another important individual in the client's life. In the question posed at the beginning of this section, you seem to be experiencing the second type of countertransference reaction because you have begun finding sessions unpleasant and want to avoid them, and because you are emotionally distancing yourself from your client. You are experiencing the same rejecting feelings toward your client as another person in her life displayed toward her, long ago. Understanding where your feelings are coming from may help you to get in touch with your real warm and accepting feelings for the client so that you can reestablish your previous good relationship.

## Suggested Readings on Transference
## and Countertransference

ARONSON, M., & SCHARFMAN, M. (Eds.). (1992). *Psychotherapy: The analytic approach.* Northvale, NJ: Jason Aronson.

CHESSICK, R. (1993). *Dictionary for psychotherapists: Dynamic concepts in psychotherapy.* Northvale, NJ: Jason Aronson.

GABBARD, G. (1990). *Psychodynamic psychiatry in clinical practice.* Washington, DC: American Psychiatric Press.

GIL, M. (1982). *The analysis of transference: Theory and technique* (Vol. 1). New York: International Universities Press.

GOLD, J., & NEMIAH, J. (Eds.). (1993). *Beyond transference.* Washington, DC: American Psychiatric Press.

KAHN, M. (1991). *Between therapist and client.* New York: Freeman.

ST. CLAIR, M. (1997). *Object relations and self-psychology: An introduction* (2nd ed.). Newbury Park, CA: Brooks/Cole.

URSANO, R., SONNENBERG, S., & LAZAR, S. (1991). *Concise guide to psychodynamic psychotherapy.* Washington, DC: American Psychiatric Press.

## VIOLENCE: ASSESSING RISK

*My next client appears angry and agitated, and his referral history reveals that he has been arrested for assault several times. How should I proceed with this client?*

Unfortunately, aggressive, acting-out behaviors occur more often within the context of clinical interactions than we would like to think. Statistics indicate that 17 percent of emergency room patients are violent and that 40 percent of psychiatrists are physically assaulted at least once during their careers (Tardiff, 1989). Safety should be your first consideration with potentially violent clients. For a counselor intern, keeping safe includes the following:

- Being alert to potentially threatening clients and dangerous situations
- Being aware of the ways your own behavior may either escalate or defuse a client's anger
- Taking precautions to protect yourself against physical assault

There is an increased risk of violence when you are counseling clients who:

- Are alcohol intoxicated or who have ingested drugs.
- Have disorders that include psychotic processes or organic brain dysfunction.
- Have underlying paranoid ideation.
- Are being restrained or committed involuntarily.
- Have poor impulse control.
- Are in the midst of family feuding, with other family members present.*

If your client is agitated or is becoming increasingly angry, you should adhere to the following behavioral guidelines:

- Position yourself in front of the client.
- Avoid too much eye contact.
- Do not touch the client.
- Provide extra interpersonal space.
- Speak slowly in a normal conversational tone.

---

*From *Psychiatric Interviewing: The Art of Understanding* by S. C. Shea. Copyright © 1988 by W. B. Saunders Company. Reprinted by permission.

- Carefully explain all your actions in order to avoid arousing suspicion (for example, "I am going to reach across this table to pick up the pen now").
- Make sure that the client is able to save face.
- Do not challenge the truthfulness of any of the client's statements.
- Encourage the client to verbalize feelings rather than acting them out (Gabbard, 1990; Shea, 1988).

If you find yourself about to face (1) any new client about whom you have little information, (2) a potentially aggressive client, or (3) a situation that could become dangerous, take steps to ensure your safety. Make certain that help is available and that you will be able to summon assistance quickly and easily. Seat yourself closer to the door; do not allow your client to place himself or herself between you and the door at any time. You may keep the door to the room ajar or open during the session if this seems appropriate. Do not turn your back to the client. That means the client should enter the room ahead of you at the start of the session. At the end of the session, you should exit first, but keep your body turned partially to the client so that you can observe his or her behavior at all times. Trust your intuition. If you feel that you are in danger, explain to your client that you are uncomfortable and need to leave. Notify other staff members immediately so that everyone can be kept safe and the dangerous client is not left unattended. If a client produces a weapon, leave the room immediately, even if the client is not threatening you personally, and summon help.

## Suggested Readings on Violence

BRIZER, D., & CROWNER, M. (Eds.). (1989). *Current approaches to the prediction of violence.* Washington, DC: American Psychiatric Press.

DICKSTEIN, L., & NADELSON, C. (Eds.). (1988). *Family violence: Emerging issues of a national crisis.* Washington, DC: American Psychiatric Press.

DURAND, V. (1991). *Severe behavior problems.* New York: Guilford.

PETERS, R., McMAHON, R., & QUINSEY, V. (Eds.). (1992). *Aggression and violence throughout the lifespan.* Newbury Park, CA: Sage.

ROTH, L. (1987). *Clinical treatment of the violent person.* New York: Guilford.

STREAN, H. (Ed.). (1984). *Psychoanalytic approaches with the hostile and violent patient.* Binghamton, NY: Haworth.

TARDIFF, K. (1989). *Concise guide to assessment and management of violent patients.* Washington, DC: American Psychiatric Press.

TOTH, H. (1992). *Violent men.* Washington, DC: American Psychological Association.

## CONCLUDING REMARKS

In this chapter we have attempted to provide guidelines for handling some of the professional dilemmas and problematic counseling situations you may face during your internship. We encourage you to consult with both your site supervisor and your program supervisor and to use other available resources whenever you need help in resolving a challenging problem during your internship.

# REFERENCES

BIHM, E., & LEONARD, P. (1993). Counseling persons with mental retardation and psychiatric disorders: A preliminary study of mental health counselors' perceptions. *Journal of Mental Health Counseling, 14*(2), 225–233.

BUTLER, R. (1983). An overview of research on aging and the status of gerontology today. *Milbank Memorial Fund Quarterly: Health and Society, 61*(3), 351–361.

BUTLER, R., & LEWIS, M. (1981). *Aging and mental health.* St. Louis: Mosby.

COREY, G. (1991). *Theory and practice of counseling and psychotherapy.* Pacific Grove, CA: Brooks/Cole.

COREY, G., & COREY, M. (1992). *Groups: Process and practice* (4th ed.). Pacific Grove, CA: Brooks/Cole.

DACEY, J., & TRAVERS, J. (1991). *Human development across the lifespan.* Dubuque, IA: Brown.

DUBOVSKY, S. (1988). *Concise guide to clinical psychiatry.* Washington, DC: American Psychiatric Press.

DULCAN, M., & POPPER, C. (1991). *Concise guide to child and adolescent psychiatry.* Washington, DC: American Psychiatric Press.

EISENGART, S., & FAIVER, C. (1996). Intuition in mental health counseling. *Journal of Mental Health Counseling, 18*(1), 41–52.

ERIKSON, E. (1963). *Childhood and society* (2nd ed.). New York: Norton.

FARBEROW, N., HEILIG, S., & PARAD, H. (1990). The suicide prevention center: Concepts and clinical functions. In H. Parad & L. Parad (Eds.), *Crisis intervention* (Vol. 2, pp. 251–274). Milwaukee, WI: Family Service America.

GABBARD, G. (1990). *Psychodynamic psychiatry in clinical practice.* Washington, DC: American Psychiatric Press.

GUTHEIL, T., & GABBARD, G. (1993). The concept of boundaries in clinical practice: Theoretical and risk-management decisions. *American Journal of Psychiatry, 150*(2), 188–196.

HUGHES, J., & BAKER, D. (1990). *The clinical child interview.* New York: Guilford.

KAHN, M. (1991). *Between therapist and client.* New York: Freeman.

LASOSKI, M. (1986). Reasons for low utilization of mental health services by the elderly. *Clinical Gerontologist, 5*(1–2), 1–18.

MORRISON, J. (1997). *When psychological problems mask medical disorders.* New York: Guilford.

RUBIN, S., & NIEMEIER, D. (1993). Non-verbal affective communication as a factor in psychotherapy. *Psychotherapy, 29*(4), 596–602.

SHEA, S. (1988). *Psychiatric interviewing: The art of understanding.* Philadelphia: Saunders.

TARDIFF, K. (1989). *Concise guide to assessment and management of violent patients.* Washington, DC: American Psychiatric Press.

TURNER, J., & HELMS, D. (1991). *Lifespan development* (4th ed.). Chicago: Holt, Rinehart and Winston.

URSANO, R., SONNENBERG, S., & LAZAR, S. (1991). *Concise guide to psychodynamic psychotherapy.* Washington, DC: American Psychiatric Press.

YALOM, I. (1985). *Theory and practice of group psychotherapy.* New York: Basic Books.

ZIVIAN, M., LARSEN, W., KNOX, V., GEKOSKI, W., & HATCHETTE, V. (1993). Psychotherapy for the elderly: Psychotherapists' preferences. *Psychotherapy, 29*(4), 668–674.

# BIBLIOGRAPHY

ARONSON, A., & SCHARFMAN, M. (1992). *Psychotherapy: The analytic approach*. Northvale, NJ: Jason Aronson.

GITLIN, M. (1990). *The psychotherapist's guide to psychopharmacology*. New York: Macmillan.

HAMILTON, G. (1992). *Self and others: Object relations theory in practice*. Northvale, NJ: Jason Aronson.

JORDAN, J., KAPLAN, A., MILLER, J., STIVER, I., & SURREY, J. (1991). *Women's growth in connection*. New York: Guilford.

LERNER, H. (1988). *Women in therapy*. Northvale, NJ: Jason Aronson.

MADANES, C. (1991). *Strategic family therapy*. San Francisco: Jossey-Bass.

SCHARFF, D. (1992). *Refinding the object and reclaiming the self*. Northvale, NJ: Jason Aronson.

SCHARFF, D., & SCHARFF, J. (1992). *Scharff notes: A primer of object relations theory*. Northvale, NJ: Jason Aronson.

THOMPSON, C., & RUDOLPH, L. (1992). *Counseling children* (3rd ed.). Pacific Grove, CA: Brooks/Cole.

# 9

# Along the Way

## A Counselor Self-Assessment

We are in the business of introspection, reflection, analysis, synthesis, and personal action. We personify what Freud (1943) called the *talking cure:* our words are our product. It certainly behooves us to look at ourselves as we move through the process of creative self-examination with our clients. Although this self-examination can take considerable courage, it can also help us appreciate how our clients may feel as they expose themselves to us in the therapeutic milieu. We have noticed that, as counselors, we sometimes become desensitized and myopic to the justifiable defenses evident as a client self-discloses issues and concerns, often at our prompting. Our own thorough self-examination may foster increased capacity for empathy and personal and professional refinement. Our own issues and countertransferences may surface in this process, in addition to other possible character quirks (perhaps even our own neuroses). All personal issues are worthy of inspection. For this reason, we recommend personal therapy for all counselors as an adjunct to training. It helps to see the process from the other side.

Dowling (1984) indicates that students can accurately evaluate themselves as well as their peers. Further, Bernard and Goodyear (1992) advocate that supervisors teach interns the principles of both peer and self-evaluation and expect such an examination before each supervisory session.

Corey and Corey (1998) likewise advise that counselors-in-training perform self-exploration so as to better understand their clients' issues. Moreover, they note that for many students, self-exploration is a more difficult task than that of apprehending their clients' situations. Benjamin (1987), who himself conquered a major obstacle—his blindness—shares his personal philosophy of self-knowledge, which involves trust in our own feelings, ideas, and intuitions. Further, as we become at ease with personal self-exploration, we not only nurture our own personal growth but also enhance our understanding of clients.

In this chapter we offer the beginnings of the self-examination process for counselors-in-training. Our purpose is to suggest areas of personal reflection and inner dialogue. Certainly, who we are as people influences our choice of profession and how we conduct ourselves as professionals. Thus we first ask you to look at yourself from a personal vantage point, examining your values, beliefs, needs, priorities, characteristics, and biases. Next assess your professional role, your career goals, and your personal traits that have influenced your career choice. Certainly we realize that it's not possible to divorce the personal from the professional; in fact, if we are fortunate, our career is an important extension of our personality. However, for the sake of clarity, we posit a somewhat artificial division. The chapter presents questions in an attempt to elicit some inner probing. We do, however, advise you to seek out trusted fellow students or professors for pertinent discussion, feedback, and validation.

## WHO AM I AS A PERSON?

1. How do I assess my developmental history up to this point of my life? What were the high and low points?

2. When did I realize I was an adult? How did I handle this realization?

3. What are my five best qualities?

4. What five areas of my life do I need to improve?

5. If an interviewer asked me what my basic philosophy of life is, what would I answer?

6. Is my glass of water half full or half empty? Why?

7. What pervasive mood do I find myself in most of the time?

8. What do I think about people in general?

9. What role do my religion, culture, ethnic values, gender, and sexual orientation play in my view of life?

10. How do I respond to those who are different from me? How open am I to multiculturality?

11. On the Meyers Briggs Type Inventory, what type am I?

12. Who are my heroes?

13. Who is the most creative person I know? Why? How can I nurture creativity in my life?

14. What is maturity?

15. What are my personal goals and objectives?

16. Where do I lie on the continuum of cooperation versus competition?
17. Who and what influenced my life?
18. What would my best friend say about me?
19. What is the biggest criticism people have of me?
20. What three adjectives best describe me?
21. Have I examined what Jung (1980) calls my "shadow"? What do I see in it?
22. Am I open to my own potential needs for counseling?
23. How do I ground and center myself?
24. Do I rely on my intuition? Why or why not? How do I rely on it?
25. Do I take time to play? How do I play?
26. What are my spiritual beliefs? How do they influence me as a person?

## WHO AM I AS A PROFESSIONAL?

1. What are my reasons for becoming a counselor?
2. Do I feel my emotional issues will be addressed and resolved by becoming a counselor?
3. What is my need to be a counselor?
4. What makes me think that I will be an effective counselor?
5. What are my countertransference issues? How do I handle my own issues when they emerge?
6. What do I expect from clients?
7. What do I expect from my profession?
8. What do I anticipate getting from colleagues?
9. What are my professional strengths and weaknesses?
10. What are my professional goals and objectives?
11. What would my fellow students say about me?
12. What would my professors and supervisors say about me?
13. With what type of clients do I wish to work? Why?
14. Do I know when and where to refer?
15. How would I handle stress and burnout?
16. How do I handle praise and criticism of my work?
17. Do I fully attend to my clients? How do I fully attend to my clients?

Every time you interact with your clients, your professional and personal attributes and self become evident. As you build rapport with your clients, they come to value the way you present yourself. Clients want the assurance of your commitment to help them. The self-assessment process empowers you: the more you know about yourself, the more control you have over your life. This control can be passed along to clients. Refinements come about through continued interaction with colleagues, clients, and others with whom we have daily contact.

Be careful not to expect too much of yourself too early in your career. For this reason we suggest being gentle with yourself; let the natural process of your professional development follow its course. Be yourself. Do not attempt to play a role or Jung's (1980) concept of a persona. Confidence should build over time. Your self-assessment enhances the formal skills, techniques, and theories you have learned in this evolving process. Finally, knowing what you can and cannot do for your clients helps you grow and strengthens your knowledge base. The ACA Code of Ethics and Standards of Practice (1995) encourages you to transfer clients to another professional if you determine that you cannot assist your clients. You cannot be everything to everybody!

Being guide and mentor—and perceived rescuer—is not an easy task. The journey of life is not an easy one; nor is it fair. Our clients know this. At times it may be difficult for us as well as our clients to maintain a positive attitude toward humanity and even existence itself. Yet we must not view life's journey as tedious, threatening, and dismal. We recall the words of a former colleague: "All we do in this lifetime is rearrange deck chairs on the Titanic." Is there more than this? We believe so. We have, however, no empirical proof, only a conviction that humanity is basically good and that we have some freedom to intervene in the condition of humankind, albeit limited. Maintaining a personal attitude of hope fosters hope in our clients; exhibiting a basic trust in humankind can engender a sense of inward and profound trust in our clients; maintaining expectations for improvement serves as impetus for change; and demonstrating positive regard for those with whom we come in contact affirms and sustains them in their humanity and dignity. It is these beliefs for which we stand.

## REFERENCES

BENJAMIN, A. (1987). *The helping interview with case illustrations*. Boston: Houghton Mifflin.

BERNARD, J., & GOODYEAR, R. (1992). Clinical evaluation. In S. Badger (Ed.), *Fundamentals of clinical supervision* (pp. 105–130). Boston: Allyn & Bacon.

COREY, M., & COREY, J. (1998). *Becoming a helper* (3rd ed.). Pacific Grove, CA: Brooks/Cole.

DOWLING, S. (1984). Clinical evaluation: A comparison of self, self with videotape, peers, and supervisors. *The Clinical Supervisor, 2*, 71–78.

FREUD, S. (1943). *A general introduction to psychoanalysis*. Garden City, NY: Garden City Publishing.

JUNG, C. (1980). *The archetypes and the collective unconscious* (R. Hull, Trans.). Princeton, NJ: Princeton University Press. (Original work published 1902.)

# BIBLIOGRAPHY

ATKINSON, D., WORTHINGTON, R., DANA, D., & GOOD, G. (1991). Etiology, beliefs, preferences for counseling orientations and counseling effectiveness. *Journal of Counseling Psychology, 38*(3), 258–264.

BALSAM, M., & BALSAM, A. (1984). *Becoming a psychotherapist.* Chicago: University of Chicago Press.

BRADSHAW, J., (1992). *Homecoming: Reclaiming and championing your inner child.* New York: Bantam.

CAMERON, J. (1992). *The artist's way: A spiritual path to higher creativity.* New York: Putnam.

CLINESS, D. (1996). *The journey of life.* Poland, OH: Banausic.

COREY, G., & COREY, M. (1997). *I never knew I had a choice.* (6th ed.). Pacific Grove, CA: Brooks/Cole.

DAY, L. (1996). *Practical intuition: How to harness the power of your instinct and make it work for you.* Scranton, PA: Harper Audio.

EMERY, M. (1994). *Intuition workbook.* Englewood Cliffs, NJ: Prentice Hall.

FULGHUM, R. (1993). *All I really need to know I learned in kindergarten: Uncommon thoughts on uncommon things.* Westminster, MD: Faucet.

FULGHUM, R. (1996). *From beginning to end: The rituals of our lives.* Westminster, MD: Faucet.

LANKTON, S., & LANKTON, C. (1983). *The answer within: A clinical framework of Erickson hypnotherapy.* New York: Brunner/Mazel.

MOORE, T. (1994). *Care of the soul: A guide to cultivating depth and sacredness in everyday life.* New York: Harperperennial.

MOORE, T. (1994). *Soul mates: Honoring the mysteries of love and relationship.* New York: Harperperennial.

MOORE, T. (1995). *Meditations: On the monk who dwells in daily life.* New York: Harperperennial.

PECK, M. (1978). *The road less traveled.* New York: Simon and Schuster.

PECK, M. (1997). *The road less traveled and beyond: Spiritual growth in an age of anxiety.* Old Tappan, NJ: Thorndike.

SUSSMAN, M. (1992). *A curious calling: Unconscious motivations for practicing psychotherapy.* Northvale, NJ: Jason Aronson.

TRACEY, T. (1991). The structure of control and influence in counseling and psychotherapy: A comparison of several definitions and measures. *Journal of Counseling Psychology, 38*(3), 265–278.

VACHON, D., & AGRESTI, A. (1992). A training proposal to help mental health professionals clarify and manage implicit values in the counseling process. *Professional Psychology: Research and Practice, 23*(6), 509–514.

# 10

# Ethical Issues

## ETHICAL PRACTICE IN COUNSELING

The term *ethical* means that a behavior or practice conforms to the specialized rules and standards for proper conduct of a particular profession. Ethics, therefore, is "a fundamental and defining aspect of professionalism" (Gibson & Pope, 1993, p. 330). The code of ethics is basic to the identity of any professional organization because it provides behavioral norms and guidelines that structure each member's professional activities as well as ensure uniformly high professional standards among all members.

Counselors-in-training have an obligation to know and to adhere to the Ethical Standards of the American Counseling Association (1995) in addition to any pertinent state codes and laws. These standards spell out distinct behavioral parameters within which all counselors, including yourself as a counselor intern, must operate. The ACA Code of Ethics is divided into eight sections, each describing one aspect of proper professional behavior; see Appendix E for the complete ACA Code of Ethics. In addition, Appendix E includes the ACA's Standards of Practice, which provide "minimal behavioral statements" summarizing the Code of Ethics and sections on policies and procedures for responding to members' requests for code interpretations and for processing complaints.

Counselor interns also need to be aware of and to function within the boundaries of their agency or institutional policies, provided that these policies do not conflict with the ACA Code of Ethics. If you perceive a possible conflict, realize that the Code of Ethics takes precedence over agency or institutional policies. We suggest that you discuss the issue with your agency supervisor and university supervisor to try to resolve the problem. If the issue is not resolved at this level, then you may wish to contact the ACA or the ethics committee of your state licensure or certification board or your state division of the ACA for further advice or assistance. Finally, as an intern and future professional counselor, you should always conduct your counseling activities in accordance with state and local laws and regulations. These rules can provide you with important professional backup and a sense of security in dealing with uncomfortable or equivocal professional situations.

## ETHICAL DILEMMAS

Ethical codes provide guidelines and principles for ethical conduct. Realistically, though, as counselors we will encounter many professional situations that are not specifically addressed by the ACA Code of Ethics (Corey, 1996; Corey, Corey, & Callanan, 1998; Gibson & Pope, 1993). Furthermore, certain clinical circumstances may involve conflicting interests or multiple ethical principles that leave us with exceedingly difficult choices. "It is the translation of a code's principles into practical directions for conduct that is the greatest challenge for most of us" (Gibson & Pope, 1993, p. 330).

For example, professional dilemmas frequently occur when personal morality conflicts with professional ethics. Our client may reveal to us that he or she is engaging in some behavior that we personally believe to be wrong or even immoral, yet we are bound by our ethical guidelines and the rules of professional confidentiality and cannot take any action outside of the treatment situation unless harm is threatened to self or others. The result can be uncomfortable cognitive dissonance. One counselor intern agonized over a client who reported driving drunk several nights a week. Although he did encourage his client to work on this issue during their sessions, the intern told us, "If I could just be a private citizen and not a professional counselor, I would make a telephone call to report this person immediately! I feel a moral responsibility to keep her off the road."

Another area of potential professional quandary involves choice of "client." For example, if you feel that a client is antisocial, presents a possibility of harm to others, or poses an outright danger to society, is that person your client or is the system or the general public your client? Where does your responsibility lie? One counselor intern described her dilemma when a client who was HIV positive disclosed that his partners were unaware of this test result and that he frequently did not adhere to safe sex practices. This intern wondered about her duty to warn and protect the partners versus her duty to protect client confidentiality. Another counselor intern described a situation in which she was counseling an undergraduate student with a history of violent, aggressive behavior, which was unknown to university administrators.

This undergraduate was granted permission to live in a dormitory on campus. The counselor intern worried about the safety of the other students, yet felt that she could not warn university officials because the client had made no specific threats to hurt anyone.

## GETTING HELP IN RESOLVING ETHICAL PROBLEMS

During your internship, and throughout your counseling career, you will grapple with difficult decisions as you try to apply the ACA Code of Ethics to ambiguous or complex counseling situations. We recommend using a variety of resources to help resolve your concerns, including supervisors and colleagues, the ACA Ethics Committee, state licensure or certification boards, and professional journals and texts that address ethics in counseling. In their book *Issues and Ethics in the Helping Professions,* Corey, Corey, and Callanan (1998) analyze many ethical dilemmas counseling professionals may confront. We encourage you to read this text thoroughly to help prevent violation of accepted practice standards and to refer to it in any questionable professional circumstances, as well.

As a counseling professional, you will find that your own values and morals, as well as your personal conceptualization of ethical counseling behavior, influence your interpretation and application of the ACA Code of Ethics. Welfel and Kitchener (1992) suggest that "the limitations of professional codes can be addressed by considering more fundamental ethical principles" (p. 180). The work of Welfel and Kitchener may provide a framework as you strive to make ethical decisions in counseling. They write that the following five basic principles underlie ethical codes in the helping professions:

1. Benefit others (implying a responsibility to do good).
2. Do no harm to others.
3. Respect the autonomy of others (including freedom of thought and freedom of action).
4. Act in a fair and just manner (meaning that the rights and interests of one individual must be balanced against the rights and interests of others).
5. Be faithful (implying trustworthiness, loyalty, and keeping promises).

Keeping these factors in mind may clarify your understanding of those ethical dilemmas in counseling where the ACA Code of Ethics does not seem to provide definitive answers.

## A BRIEF SUMMARY OF THE ACA CODE OF ETHICS

The first step in upholding the ethical standards and practices of the counseling profession during your internship is gaining a working knowledge of the ACA Code of Ethics. Many counseling students tend to avoid thoroughly reading or examining the ACA Code of Ethics. They feel that it is too complicated, too lengthy, and inaccessible because it is often relegated to an appendix at the back of a book rather than being included within the main text. In addition, the ACA Code of Ethics may seem somewhat remote from their immediate needs and frame of reference as graduate students rather than as counseling practitioners. However, as a counselor intern, your role has evolved from that of primarily a student to that of essentially a clinician. We have therefore provided the following summary of the ACA Code of Ethics in an effort to make it more "user friendly." It is interesting to note that the code is grounded in the eight CACREP (Council on Accreditation of Counseling and Related Educational Programs) core areas of study. We strongly recommend that you set aside the time, early in your internship, to study the ACA Code of Ethics in its original, unabbreviated form and keep a copy handy for reference. A solid understanding of the ACA Code of Ethics will provide you with a knowledge base as you struggle to apply ethical principles to the diversity of challenging situations you will encounter during your counseling internship and your counseling career.

## Section A: The Counseling Relationship

Counselors always respect the rights, the freedom, the integrity, and the welfare of the client, whether in an individual relationship or in a group setting. Counselors believe that clients have the potential to grow and change and strive to empower them in this endeavor. You will develop with clients measurable, observable, and realistic treatment plans that provide considerable flexibility in the therapeutic process. Counselors value the importance of the family as both a support system and a resource for clients, while always protecting the individual privacy rights of their clients. Counselors need to understand the strengths and limitations of their clients when career and employment needs become factors in the treatment process. Clients' multicultural diversity must always be respected and understood; at the same time counselors must gain insight into their own unique identities in relating to clients. Counselors explain the goals, structure, expectations, rules, treatment techniques, proposed treatment outcomes, and any limitations of counseling, including professional competencies and any limitations, to clients at the outset of the counseling relationship. Clients (or guardians or parents of minors) can terminate the counseling relationship if they perceive no interest and benefit. Counselors do not enter a professional relationship with a client who is already involved in another counseling relationship unless proper notification is given and a request to terminate and transfer is initiated. Counselors maintain professional boundaries and do not make their personal needs an issue. Dual relationships (such as both counselor and teacher, supervisor and supervisee, or counselor and friend) should be avoided, and referral should be made. Counselors view sexual intimacies or relationships with clients and former clients as inappropriate and unethical. Boundaries must be clear and roles defined so as to avoid any conflict. Counselors assume responsibility for protecting group members from psychological or physical suffering resulting from participation in the group. Counselors protect group members by screening prospective members, by maintaining an awareness of group interactions and relationships, and by making available professional assistance for members who require such intervention, both during and after group sessions. Group members need to be apprised of the goals and objectives, rules, roles, and projected treatment outcomes. Counselors advise group members about respecting rules of confidentiality and avoid the dual roles of performing both individual and group counseling. Counselors assist clients with necessary financial arrangements related to the counseling relationship, realizing that fees are an integral element of the therapeutic process. When counselors have to interrupt their service delivery to clients, they must ensure the continuity of treatment. In addition, counselors advise clients when they are unable to be of professional help and then assist these clients in finding alternatives. When using computers as part of counseling services, counselors take precautions to make certain that the client is intellectually, physically, and emotionally capable of using a computer, that the computer program is appropriate and understandable, and that follow-up counseling is provided. Counselors who design computer software ensure that self-help/stand-alone software has actually been designed as such. Counselors designing such software also provide descriptions of expected outcomes, suggestions for use, inappropriate applications, possible benefits, and a manual defining any other essential data. Computer programs, methods, and techniques adhere to applicable professional standards, ethics, and guidelines.

## Section B: Confidentiality

Counselors maintain the standard of confidentiality as they work to establish the therapeutic relationship. This standard is violated only when the safety and well-being of the clients, those individuals around them, and the general community are in jeopardy. Counselors inform clients of the limitations of confidentiality and identify situations when confidentiality may be breached, including court orders, treatment team approaches, clinical and fiscal audits, and quality assurance reviews. Confidentiality is preserved when working with groups, families, and minor or incompetent clients. Client records meet the standards in order to safeguard the integrity of the individual. Clients have the right to access their records with the proviso that they are competent in understanding their records or have an advocate to interpret the content. Counselors who use client data for research purposes ensure client anonymity. Counselors who are consultants share client information only with individuals who are clearly concerned with and responsible for the case. Individuals must sign releases of information when records or other client data are shared in any form.

## Section C: Professional Responsibility

Counselors operate within their scope of competencies, education and training, and professional experience. Counselors are committed to continuing education and awareness in working with diverse populations, diagnoses, and theories and techniques of treatment. Counselors accept employment opportunities that are congruent with their education and training and continually solicit feedback about the quality of their professional work. Counselors seek the guidance and input of other colleagues when questioning ethical obligations or concerns of professional practice. Counselors avoid offering or accepting counseling responsibilities when impaired either physically or emotionally. Counselors accurately present their credentials and scope of practice to the general public and their clients. They rely on the rules and guidelines of their respective licensure or certification boards. Counselors immediately correct any errors in education, training, licensure, or certification. Counselors provide their services to individuals regardless of age, color, culture, disability, ethnic group, gender, race, religion, sexual orientation, or socioeconomic status. Moreover, counselors respect differences in colleagues' therapeutic approaches. They endeavor to take into account the varying traditions and practices of other professional groups with whom they work.

## Section D: Relationships with Other Professionals

Counselors define and describe their professional roles and responsibilities with employers, employees, colleagues, and supervisors. These delineations are reflected in working agreements, clinical relationships, and issues of confidentiality; they also adhere to professional standards. Peer review of the counselor's quality of work, as reflected in ongoing professional evaluations, provides professional feedback that improves service delivery. Continuing in-service education is central to the refinement of clinical knowledge and skills. Counselors' personal and professional practices demonstrate respect for the rights and well-being of colleagues and at the same time maintain the highest level of professional service. Counselors in administrative or supervisory capacities ensure that staff operate within their level of competency. Counselors maintain the highest standard of professional conduct in their relationships with both clients and affiliated agencies. Counselors abide by the policies and procedures of their agencies. In addition, counselors understand the need for ethical reassessment and modification of policies and procedures in order to promote better service.

## Section E: Evaluation, Assessment, and Interpretation

Counselors use educational and psychological assessment techniques in the best interests of client welfare. Counselors do not misuse test results and interpretations and endeavor to prevent others from doing so. Counselors strive to respect clients' rights to know test results, interpretations, and the basis for conclusions and recommendations. Counselors are aware of limits of competency, performing only those testing and assessment services in which they have been adequately trained. Counselors consider reliability and validity data; the currency or obsolescence of tests; proper administration of any technique utilized; and awareness of socioeconomic, cultural, and ethnic factors. Conoley and Impara's (1998) *The Thirteenth Mental Measurement Yearbook* and *Tests in Print IV* (1994) are excellent resources for understanding standardized tests. Counselors are responsible for ensuring the appropriate application, scoring, interpretation, and use of assessment instruments. Before administering an assessment instrument, counselors clearly explain the nature and purpose of the assessment as well as how the results will be used. Counselors are careful in providing the proper diagnoses. Counselors are careful in conducting clinical interviews and selecting assessment techniques. Counselors strive to maintain the integrity and security of tests and other assessment instruments in compliance with legal and contractual requirements.

## Section F: Teaching, Training, and Supervision

Counselors responsible for training counselor education students and counselor supervisees are competent practitioners and teachers. They model the highest ethical standards of the profession, maintaining appropriate professional and social boundaries. Counselors are adequately trained in supervision methods. Further, they ensure that students and supervisees assume a professional role in providing services. Counselor educators advise prospective students concerning program requirements, including the scope of professional skills development. They provide a comprehensive view of the counseling profession, including employment opportunities, before admission to the counselor education training program. Counselors establish training sites that integrate academic study and supervised practice. These sites expose trainees to diverse theoretical orientations and populations. This exposure affords the trainees the opportunity to increase their competencies and knowledge base. Students and supervisees are provided clear guidelines pertaining to competency levels expected and subsequent appraisal and evaluation methods for both didactic and experiential training components. Periodic performance appraisal with feedback is provided. Counselors communicate expectations of trainees' roles and responsibilities to their trainees. Care is taken in confirming that site supervisors are qualified to provide supervision, including the ethical responsibilities of this role. Counselor educators must avoid duplicity of roles as field supervisor and training program supervisor. At no time should their supervisor counsel students or supervisees; in this case, referral is made to another practitioner.

## Section G: Research and Publication

Counselors performing research with human subjects adhere to ethical principles and regulations consistent with federal and state laws and scientific standards. At all times counselors protect the rights of research participants. Counselors avoid anything that could injure their subjects. The principal researcher holds the primary responsibility for ensuring the maintenance of ethical practice. However, all counselors involved in a study are aware of ethical principles and assume responsibility for their actions. Counselors inform subjects of the purpose, procedures, risks, and benefits of the investigation if possible. Researchers carefully disguise subjects' identities. Participant data are kept in strictest confidence. Further, counselors honor all commitments to subjects. Counselors report research findings in an ethical manner, noting any conditions that may misleadingly affect results. Original research data are reported in order to allow for replication. Previous research findings are appropriately and objectively cited. Counselors publishing material are careful to credit the work of others, such as through the use of citations, references, footnotes, acknowledgments, or joint authorship. Counselors jointly performing research are responsible for the accuracy and thoroughness of the information they contribute. In addition, counselors who collaborate incur an ethical responsibility to other collaborators to honor agreed-upon commitments. Counselors do not submit work to more than one journal or publisher at a time. Counselors obtain publisher permission for reprinting manuscripts.

### Section H: Resolving Ethical Issues

Counselors are familiar with the Code of Ethics, realizing that ignorance or misunderstanding of an ethical responsibility is not an acceptable defense against a charge of unethical conduct. When counselors have specific ethical questions, they confer with other counselors or appropriate authorities. When ethical issues emerge regarding their affiliated organization, counselors specify the nature of such conflict, initially attempting to resolve the issue with their supervisor or other responsible official. Counselors may hypothetically discuss ethical issues and concerns with colleagues. Counselors take cautions so as not to initiate, participate in, or encourage the filing of unwarranted or harmful ethical complaints. Counselors cooperate in any proceedings to investigate ethical complaints at all levels.

# A MODEL FOR ETHICAL COUNSELING PRACTICE

Ethical behavior requires more than a familiarity with the profession's code of ethics. Often implementing ethical decisions with ethical action involves substantial personal and professional risk or discomfort (Gawthrop & Uhleman, 1992). The motivation to accept these possible consequences to self of ethical actions may be considered a moral decision. Welfel and Kitchener (1992) present a model for moral behavior, which consists of four components, each a psychological process. They consider this model from the perspective of the professional helper, and we suggest that their ideas have great merit for you as a counselor intern and for all professional counselors who are concerned with ethical practice.

The components of moral and ethical behavior described by Welfel and Kitchener (1992) include the following:

1. Interpreting the situation as a moral one (recognizing the existence of an ethical problem and acknowledging that its outcome affects the welfare of your client and other people).

2. Using moral reasoning (understanding and applying ethical codes; using ethical principles where applicable).

3. Deciding what one intends to do (having recognized that an ethical problem exists, and having decided what the ethical response might be, the professional decides whether to proceed despite the discomfort of confrontation or other ethical action).

4. Implementing moral action (persevering with ethical behavior despite costs to self or great pressure to withdraw; for example, being steadfast in the face of legal action or disapproval from colleagues).

## CONCLUDING REMARKS

You will surely confront a broad scope of ethical issues and dilemmas during your counseling internship and throughout your professional career. A knowledge of the ACA Code of Ethics gives you a foundation for your ethical decisions; however, you will need to examine your own values closely as you search for acceptable solutions to some of these ethical problems. Herlihy and Corey (1996) view the Code of Ethics as a necessary and vital core component of the education and training of counselors, which includes sound professional conduct and accountability and which results in improved practice. In addition, they feel ethical codes provide a combination of rules and utilitarian principles in ways that require interpretation. We encourage you to discuss ethical questions and issues with your colleagues, supervisors, and professors and to refer to the professional journals and books that address counseling ethics as you search for those answers.

## REFERENCES

CONOLEY, J.C., & IMPARA, J.C. (Eds.). (1998). *The thirteenth mental measurements yearbook.* Lincoln, Nebraska: The Buros Institute of Mental Measurements, The University of Nebraska.

COREY, G. (1996). *Theory and practice of counseling and psychotherapy* (5th ed.). Pacific Grove, CA: Brooks/Cole.

COREY, G., COREY, M., & CALLANAN, P. (1998). *Issues and ethics in the helping professions* (5th ed.). Pacific Grove, CA: Brooks/Cole.

ETHICAL STANDARDS OF THE AMERICAN COUNSELING ASSOCIATION (1995). Alexandria, VA: American Association of Counseling and Development Governing Council.

GAWTHROP, J., & UHLEMAN, M. (1992). Effects of the problem-solving approach in ethics training. *Professional Psychology: Research and Practice, 23*(1), 38–42.

GIBSON, W., & POPE, K. (1993). The ethics of counseling: A national survey of certified counselors. *Journal of Counseling and Development, 71*(3), 330–336.

HERLIHY, B., & COREY, G. (1996). *ACA ethical standards casebook.* Alexandria, VA: American Counseling Association.

MURPHY, L.L., CONOLEY, J.C., & IMPARA, J.C. (eds.). (1994). *Tests in print IV.* Volume II. Lincoln, Nebraska: The Buros Institute of Mental Measurements, The University of Nebraska.

WELFEL, E., & KITCHENER, K. (1992). Introduction to the special section: An agenda for the 90's. *Professional Psychology: Research and Practice, 23*(3), 179–181.

# BIBLIOGRAPHY

ARTHUR, G., & SWANSON, C. (1993). *Confidentiality and privileged communication.* Alexandria, VA: American Counseling Association.

BULLIS, R. (1992). *Law and management of a counseling agency or private practice.* Alexandria, VA: American Counseling Association.

BURKE, M., & MIRANTI, J. (1992). *Ethical and spiritual values in counseling.* Alexandria, VA: American Counseling Association.

CONOLEY, J.C., & IMPARA, J.C. (Eds.). (1998). *The thirteenth mental measurements yearbook.* Lincoln, Nebraska: The Buros Institute of Mental Measurements, The University of Nebraska.

COREY, G. (1996). *Student manual for Corey's theory and practice of counseling and psychotherapy* (5th ed.). Pacific Grove, CA: Brooks/Cole.

GABBARD, G. (Ed.). (1989). *Sexual exploitation in professional relationships.* Washington, DC: American Psychiatric Press.

GOODYEAR, R., CREGO, C., & JOHNSTON, M. (1992). Ethical issues in the supervision of student research: A study of critical incidents. *Professional Psychology: Research and Practice, 23*(2), 203–210.

GRESSARD, C.F., & KEEL, L. (1998). An introduction to the ACA code of ethics and standards of practice. *Counseling Today, 40*(8), 16–23.

GUTHEIL, T., & GABBARD, G. (1993). The concept of boundaries in clinical practice: Theoretical and risk-management dimensions. *American Journal of Psychiatry, 150*(2), 188–196.

HARDING, A., GRAY, L., & NEAL, M. (1993). Confidentiality limits with clients who have HIV: A review of ethical and legal guidelines and professional policies. *Journal of Counseling and Development, 71*(3), 297–305.

HASS, L.J., & MALOUF, J.L. (1995). *Keeping up the good work: A practitioner's guide to mental health ethics* (2nd ed.). Sarasota, L: Professional Resource Press.

HERLIHY, B., & GOLDEN, L. (1990). *Ethical standards casebook* (4th ed.). Alexandria, VA: American Counseling Association.

HOPKINS, B., & ANDERSON, B. (1990). *The counselor and the law* (3rd ed.). Alexandria, VA: American Counseling Association.

HOPKINS, W.E. (1997). *Ethical dimensions of diversity.* Thousand Oaks, CA: Sage.

KITCHENER, K. (1992). Psychologist as teacher and mentor: Affirming ethical values throughout the curriculum. *Professional Psychology: Research and Practice, 70*(3), 190–195.

MAcNAIR, R. (1992). Ethical dilemmas of child abuse reporting: Implications for mental health counselors. *Journal of Mental Health Counseling, 14*(2), 127–136.

MITCHELL, R. (1991). *Documentation in counseling records.* Alexandria, VA: American Counseling Association.

MURPHY, L.L., CONOLEY, J.C., & IMPARA, J.C. (eds.). (1994). *Tests in print IV.* Volume II. Lincoln, Nebraska: The Buros Institute of Mental Measurements, The University of Nebraska.

NATIONAL BOARD FOR CERTIFIED COUNSELORS ETHICAL CODE. (1998). *National Board for Certified Counselors News Notes. 14*(3).

PEDERSEN, P. (1995). Culture-centered ethical guidelines for counselors. In J.G. Ponterotto, J.M. Casas, L.A. Suzuki, & C.M. Alexander (Eds.), *Handbook of multicultural counseling* (pp. 34–49). Thousand Oaks, CA: Sage.

PEDERSEN, P. (1997). The cultural context of the American counseling association code of ethics. *Journal of Counseling and Development, 76*(1), 23–28.

POPE, K., & VETLER, V. (1992). Ethical dilemmas encountered by members of the American Psychological Association: A national survey. *American Psychologist, 47*(3), 397–411.

ROSENBAUM, M. (Ed.). (1982). *Ethics and values in psychotherapy: A guidebook.* New York: Free Press.

RUTTER, P. (1989). *Sex in the forbidden zone*. New York: Fawcett.

SALO, M., & SHUMATE, S. (1993). *Counseling minor clients*. Alexandria, VA: American Counseling Association.

STEVENS-SMITH, R., & HUGHES, M. (1993). *Legal issues in marriage and family counseling*. Alexandria, VA: American Counseling Association.

VASQUEZ, M. (1992). Psychologist as clinical supervisor: Promoting ethical practice. *Professional Psychology: Research and Practice, 70*(2), 196–202.

WELFEL, E. (1992). Psychologist as ethics educator: Successes, failures, and unanswered questions. *Professional Psychology: Research and Practice, 70*(3), 182–189.

WELFEL, E. (1998). *Ethics in counseling and psychotherapy: Standards, research, and emerging issues*. Pacific Grove, CA: Brooks/Cole.

# 11

# Finishing Up

W e anticipate that it is with great pride and a sense of accomplishment that you now approach the end of your placement. We hope that your journey in clinical growth has been a fruitful one and that you are now ready to enter the "real world" of therapy.

However, several items need attention before you exit this preprofessional experience. We have compiled a checklist of activities pertaining to clients, services, their continuity of care, and programmatic issues that will help you finish up the process.

### Client Transfer

- Determine when you will notify your clients of the impending completion of your internship.
- Address clients' feelings and provide opportunities to say good-bye.
- Arrange for a new counselor to be assigned.
- Schedule a counseling session for your clients to be introduced to their new counselor.
- Decide with your clients the stage of your counseling relationship conducive for the transfer to occur.
- Maintain precise documentation permitting the new counselor and client to know exactly where to begin services (treatment plan, transfer summary, progress notes).

- Notify the social service agencies that provided case management/advocacy of your impending departure and who will replace you.

- Clarify the current status of issues you have been addressing with various outside agencies.

- Verify that all letters, reports, and records that are requested or needed are sent.

- Have the client sign new releases of information so exchange of information continues.

## Terminating Clients

- Attempt to achieve the goals of treatment plans for clients to be terminated.

- Notify clients of services available to them if the need arises (such as support groups, crisis telephone help lines, and emergency counseling services).

- Adhere to the detailed provisions of after-care services (regarding medication, review appointments, support groups, bibliotherapy, and so on).

## Program Coverage

- Arrange for coverage for groups.

- Remove yourself from schedules of intake assessments, emergency services, didactic programs, and research activities.

- Curtail all client and program services within two weeks of departure from the site.

- Offer your telephone number and address if the agency needs to contact you.

You should schedule an exit interview with your site supervisor and your university supervisor to review your performance and to complete the final evaluation. You will discuss your areas of competence (therapeutic orientation, theories, techniques), the clinical population with which you were most effective, and the service areas (individual, group, intake, emergency, case management) in which you demonstrated the most skill. General impressions and recommendations should be discussed at this time. Your input on your performance should be considered, and you should have sufficient opportunities to ask questions and get any needed clarification. Your supervisors should have been providing feedback on your performance during the entire placement; thus the final evaluation should not be a time for surprises. Your comments regarding your site supervisor's role and his or her relationship with you can be discussed. Ideally, your comments on the internship experience will be welcomed and will add to your supervisor's effectiveness.

Your site supervisor and university supervisor will probably endorse your successful completion of the internship by writing an evaluation form (see Appendix K for a sample evaluation of intern form). At the conclusion of this meeting, take the opportunity to complete any necessary licensure or certification forms.

Finally, your supervisors may recommend ways to present your experiences clearly and concisely in your resume. As a practitioner, your site supervisor understands what would increase your marketability. He or she may share information about employment opportunities and career trends. The site supervisor may suggest that you network with mental health professionals regarding job opportunities and strategies. This would also be a good time to request a professional reference.

Recognize that you have an obligation to yourself to examine your own feelings regarding this termination (refer to Appendix L for a sample intern evaluation of site and supervisor form). After all, has this internship not been a developmental process for you? Certainly your clients have moved through a therapeutic process, but you yourself have also made an educational journey consisting of several discrete steps, including starting, acclimating (or adjusting), working, assessing, integrating, and finishing.

- Starting: You bring a fund of knowledge, enthusiasm, and eagerness to learn, an openness to experience our profession in the real world.

- Adjusting: You adapt to a new environment and individuals with the goal of understanding how you fit into the overall dynamics of the agency.

- Working: You are now able to use your knowledge and skills. You aspire to establish credibility as a counselor.

- Assessing: You evaluate the quality of your performance and test your understanding of the intricacies of the counseling field. In addition, you solicit feedback regarding your efforts.

- Integrating: You are able to implement the feedback you receive, coupled with new practical knowledge gained, to become a more effective counselor.

- Finishing: You feel prepared to transfer the formal and practical knowledge into the counseling field.

We hope that this book has been helpful throughout your internship. We have tried to provide the practical information you needed to function in your role as counselor-in-training. We also hope that during your internship you were able to integrate your theoretical training with the practical expectations and demands of our profession.

We like our clients to take a neatly wrapped package away with them from therapy, a sort of gift to themselves, one that they can feel good about. This package contains much hard work, personal growth, independence, a sense of accomplishment and survival, and, in Tillich's (1952) words, "the courage to be." Likewise, we want you to take a neatly wrapped package away with you as you complete this field experience. Listed here are the contents of your internship gift to yourself. (Please permit us to wax philosophical.)

- May you have the sense that you have participated in a meaningful professional and educational experience that has touched you personally.

- Emerson once wrote that one criterion for determining success is the knowledge that "by your having lived someone has breathed easier." May you come away with a commitment to assist others in their independence from you.

- May you develop a sense of competency: competency in skills, interactions, theoretical knowledge base, and asking for information and help when you do not know the answers.

- May you continue to grow and change and adapt to new and uncertain circumstances in both your personal life and your professional life, which are, after all, inextricably intertwined.

- May you leave a positive mark on others, as we hope we have done with you.

We wish you much success and happiness in our profession.

## REFERENCES

TILLICH, P. (1952). *The courage to be.* New Haven, CT: Yale University Press.

# APPENDIX A

# Sample Résumé

**CORY A. JONES**
2300 Main Street
Cleveland, OH 44107
(216) 555-1111

| | |
|---|---|
| *Objective* | An internship in the human services and counseling area |
| *Education* | B.A., Hiram College, Hiram, OH: June 1996<br>Major: Psychology (Departmental Honors)<br>Cumulative GPA: 3.5 |
| | M.A. in Community Counseling, John Carroll<br>University, Cleveland, OH: Expected May 2002 |
| *Work experience* | *Neighboring Mental Health Center,* Mentor, OH:<br>1996–present<br>Case Manager |
| | *Cleveland Memorial Hospital,* Cleveland, OH: 1995–1996<br>Case Aide |
| *Honors and activities* | Dean's list, Hiram College: 1993–1996<br>Departmental Honors, Hiram College: 1996<br>Tutoring Program, Cleveland, OH: 1995<br>Psychology Club, Hiram College: 1994–1996<br>Copresenter, American Counseling Association: 2000 |
| *References* | Available upon request |

# APPENDIX B

# Sample Thank-You Note

Cory A. Jones
2300 Main Street
Cleveland, OH 44107
(216) 555-1111

Daniel Noday, PhD
Cleveland Counseling Agency
22000 Euclid Avenue
Cleveland, OH 44120

Dear Dr. Noday:

I appreciate your taking time to meet with me yesterday. I enjoyed having the opportunity to speak with you and to learn about your agency.

I am interested in a field placement with the Cleveland Counseling Agency beginning in September 2001. I will call you in about a week to discuss this possibility.

Thank you in advance for your consideration.

Sincerely,

Cory Jones

# APPENDIX C

# Sample Internship Announcement

INTERNSHIP/PRACTICUM PLACEMENT
FOR GRADUATE STUDENTS IN COUNSELING

Available for the Fall Semester
Cleveland Mental Health Center

Opportunities for

- Mental health counseling with a wide range of clients
- Participating in current aspects of the agency's outpatient and community education programs
- Developing new services for clients

Training supervision toward Ohio licensure in counseling available.

Contact: Marc Noday, M.D.
Clinical Director
(216) 555-2200

# APPENDIX D

# NBCC Coursework
# Requirements

## NBCC COURSEWORK AREA DESCRIPTIONS

1. Human Growth and Development includes studies that provide an understanding of the nature and needs of individuals at all developmental levels. Studies in this area include, but are not limited to, the following:

   a. Theories of individual and family development and transitions across the life span

   b. Theories of learning and personality development

   c. Human behavior, including an understanding of developmental crises, disability, addictive behavior, psychopathology, and environmental factors that affect both normal and abnormal behavior

   d. Strategies for facilitating development over the life span

   e. Ethical considerations

2. Social/Cultural Foundations include studies that provide an understanding of issues and trends in a multicultural and diverse society. Studies in this area include, but are not limited to, the following:

   a. Multicultural and pluralistic trends, including characteristics and concerns of diverse groups

   b. Attitudes and behaviors based on such factors as age, race, religious preference, physical disability, sexual orientation, ethnicity and culture, family patterns, gender, socioeconomic status, and intellectual ability

   c. Individual, family, and group strategies with diverse populations

   d. Ethical considerations

3. Helping Relationships include studies that provide an understanding of counseling and consultation processes. Studies in this area include, but are not limited to, the following:

Source: National Board for Certified Counselors. Coursework Requirements (1993). Reprinted by permission of NBCC.

a. Counseling and consultation theories, including both individual and systems perspectives as well as coverage of relevant research and factors considered in applications

b. Basic interviewing, assessment, and counseling skills

c. Counselor and consultant characteristics and behaviors that influence helping processes, including age, gender, and ethnic differences, verbal and nonverbal behaviors, and personal characteristics, orientations, and skills

d. Client or consultee characteristics and behaviors that influence helping processes, including age, gender, ethnic differences, verbal and nonverbal behaviors, and personal characteristics, orientations, and skills

4. Group Work includes studies that provide an understanding of group development, dynamics, counseling theories, group counseling methods and skills, and other group work approaches. Studies in this area include, but are not limited to, the following:

a. Principles of group dynamics, including group process components, developmental stage theories, and group members' roles and behaviors

b. Group leadership styles and approaches, including characteristics of various types of group leaders and leadership styles

c. Theories of group counseling, including commonalties, distinguishing characteristics, and pertinent research and literature

d. Group counseling methods, including group counselor orientations and behaviors, ethical standards, appropriate selection criteria, and methods of evaluating effectiveness

e. Approaches used for other types of group work, including task groups, prevention groups, support groups, and therapy groups

f. Ethical considerations

5. Career and Lifestyle Development includes studies that provide an understanding of career development and related life factors. Studies in this area include, but are not limited to, the following:

a. Career development theories and decision-making models

b. Career, avocational, educational, and labor market information resources, visual and print media, and computer-based career information systems

c. Career development program planning, organization, implementation, administration, and evaluation

d. Interrelationships among work, family, and other life roles and factors, including multicultural and gender issues as related to career development

e. Career and educational placement, follow-up, and evaluation

f. Assessment instruments and techniques relevant to career planning and decision making

    g. Computer-based career development applications and strategies, including computer-assisted career guidance systems

    h. Career counseling processes, techniques and resources, including those applicable to specific populations

    i. Ethical considerations

6. Appraisal includes studies that provide an understanding of individual and group approaches to assessment and evaluation. Studies in this area include, but are not limited to, the following:

    a. Theoretical and historical bases for assessment techniques

    b. Validity, including evidence for establishing content, construct, and empirical validity

    c. Reliability, including methods of establishing stability, internal, and equivalence reliability

    d. Appraisal methods, including environmental assessment, performance assessment, individual and group test and inventory methods, behavioral observations, and computer-managed and computer-assisted methods

    e. Psychometric statistics, including types of assessment scores, measures of central tendency, indexes of variability, standard errors, and correlations

    f. Age, gender, ethnicity, language, disability, and culture factors related to assessing and evaluating individuals and groups

    g. Strategies for selecting, administering, interpreting, and using assessment and evaluation instruments and techniques in counseling

    h. Ethical considerations

7. Research and Program Evaluation includes studies that provide an understanding of types of research methods, basic statistics, and ethical and legal consideration in research. Studies in this area include, but are not limited to, the following:

    a. Basic types of research methods, including qualitative and quantitative research designs

    b. Basic parametric and nonparametric statistics

    c. Principles, practices, and applications of needs assessment and program evaluation

    d. Uses of computers for data management and analysis

    e. Ethical and legal considerations in research

8. Professional Orientation includes studies that provide an understanding of all aspects of professional functioning, including history, roles, organizational structures, ethics, standards, and credentialing. Studies in this area include, but are not limited to, the following:

    a. History of the helping profession, including significant factors and events

b. Professional roles and functions, including similarities and differences with other types of professionals

c. Professional organizations, primarily the ACA, its dividing branches, and affiliates, including membership benefits, activities, services to members, and current emphases

d. Ethical standards of the ACA and related entities, ethical and legal issues, and their applications to various professional activities (such as appraisal, group work)

e. Professional preparation standards, their evolution, and current applications

f. Professional credentialing, including certification, licensure and accreditation practices and standards, and the effects of public policy on these issues

g. Public policy processes, including the role of the professional counselor in advocating on behalf of the profession and its clientele

9. Field Experience includes studies that provide supervised counseling experience in an appropriate work setting of at least two academic terms and taken through a regionally accredited institution. Three years of post master's counseling experience with supervision may be substituted for the first practicum.

# APPENDIX E

# American Counseling Association Code of Ethics and Standards of Practice

*(Approved by the Governing Council, April 1995)*

## PREAMBLE

The American Counseling Association is an educational, scientific, and professional organization whose members are dedicated to the enhancement of human development throughout the life span. Association members recognize diversity in our society and embrace a cross-cultural approach in support of the worth, dignity, potential, and uniqueness of each individual.

The specification of a code of ethics enables the association to clarify to current and future members, and to those served by members, the nature of the ethical responsibilities held in common by its members. As the code of ethics of the association, this document establishes principles that define the ethical behavior of association members. All members of the American Counseling Association are required to adhere to the *Code of Ethics* and the *Standards of Practice*. The *Code of Ethics* will serve as the basis for processing ethical complaints initiated against members of the association.

## CODE OF ETHICS

### Section A: The Counseling Relationship

A.1. CLIENT WELFARE

a. *Primary Responsibility.* The primary responsibility of counselors is to respect the dignity and to promote the welfare of clients.

b. *Positive Growth and Development.* Counselors encourage client growth and development in ways that foster the clients' interest and welfare; counselors avoid fostering dependent counseling relationships.

c. *Counseling Plans.* Counselors and their clients work jointly in devising integrated, individual counseling plans that offer reasonable promise of success and are consis-

ACA Code of Ethics and Standards of Practice © ACA 1995. Reprinted with permission. No further reproduction authorized without written permission of the American Counseling Association.

tent with abilities and circumstances of clients. Counselors and clients regularly review counseling plans to ensure their continued viability and effectiveness, respecting clients' freedom of choice. (See A.3.b.)

d. *Family Involvement.* Counselors recognize that families are usually important in clients' lives and strive to enlist family understanding and involvement as a positive resource when appropriate.

e. *Career and Employment Needs.* Counselors work with their clients in considering employment in jobs and circumstances that are consistent with the clients' overall abilities, vocational limitations, physical restrictions, general temperament, interest and aptitude patterns, social skills, education, general qualifications, and other relevant characteristics and needs. Counselors neither place nor participate in placing clients in positions that will result in damaging the interest and the welfare of clients, employers, or the public.

## A.2. RESPECTING DIVERSITY

a. *Nondiscrimination.* Counselors do not condone or engage in discrimination based on age, color, culture, disability, ethnic group, gender, race, religion, sexual orientation, marital status, or socioeconomic status. (See C.5.a., C.5.b., and D.1.i.)

b. *Respecting Differences.* Counselors will actively attempt to understand the diverse cultural backgrounds of the clients with whom they work. This includes, but is not limited to, learning how the counselor's own cultural/ethnic/racial identity impacts her/his values and beliefs about the counseling process. (See E.8. and F.2.I.)

## A.3. CLIENT RIGHTS

a. *Disclosure to Clients.* When counseling is initiated, and throughout the counseling process as necessary, counselors inform clients of the purposes, goals, techniques, procedures, limitations, potential risks and benefits of services to be performed, and other pertinent information. Counselors take steps to ensure that clients understand the implications of diagnosis, the intended use of tests and reports, fees, and billing arrangements. Clients have the right to expect confidentiality and to be provided with an explanation of its limitations, including supervision and/or treatment team professionals; to obtain clear information about their case records; to participate in the ongoing counseling plans; and to refuse any recommended services and be advised of the consequences of such refusal. (See E.5.a. and G.2.)

b. *Freedom of Choice.* Counselors offer clients the freedom to choose whether to enter into a counseling relationship and to determine which professional(s) will provide counseling. Restrictions that limit choices of clients are fully explained. (See A.1.c.)

c. *Inability to Give Consent.* When counseling minors or persons unable to give voluntary informed consent, counselors act in these clients' best interests. (See B.3.)

## A.4. CLIENTS SERVED BY OTHERS

If a client is receiving services from another mental health professional, counselors, with client consent, inform the professional persons already involved and develop clear agreements to avoid confusion and conflict for the client. (See C.6.c.)

## A.5. PERSONAL NEEDS AND VALUES

a. *Personal Needs.* In the counseling relationship, counselors are aware of the intimacy and responsibilities inherent in the counseling relationship, maintain respect for clients, and avoid actions that seek to meet their personal needs at the expense of clients.

b. *Personal Values.* Counselors are aware of their own values, attitudes, beliefs, and behaviors and how these apply in a diverse society and avoid imposing their values on clients. (See C.5.a.)

### A.6. DUAL RELATIONSHIPS

a. *Avoid When Possible.* Counselors are aware of their influential positions with respect to clients, and they avoid exploiting the trust and dependency of clients. Counselors make every effort to avoid dual relationships with clients that could impair professional judgment or increase the risk of harm to clients. (Examples of such relationships include, but are not limited to, familial, social, financial, business, or close personal relationships with clients.) When a dual relationship cannot be avoided, counselors take appropriate professional precautions, such as informed consent, consultation, supervision, and documentation, to ensure that judgment is not impaired and no exploitation occurs. (See F.1.b.)

b. *Superior/Subordinate Relationships.* Counselors do not accept as clients superiors or subordinates with whom they have administrative, supervisory, or evaluative relationships.

### A.7. SEXUAL INTIMACIES WITH CLIENTS

a. *Current Clients.* Counselors do not have any type of sexual intimacies with clients and do not counsel persons with whom they have had a sexual relationship.

b. *Former Clients.* Counselors do not engage in sexual intimacies with former clients within a minimum of two years after terminating the counseling relationship. Counselors who engage in such relationship after two years following termination have the responsibility to thoroughly examine and document that such relations did not have an exploitative nature, based on factors such as duration of counseling, amount of time since counseling, termination circumstances, client's personal history and mental status, adverse impact on the client, and actions by the counselor suggesting a plan to initiate a sexual relationship with the client after termination.

### A.8. MULTIPLE CLIENTS

When counselors agree to provide counseling services to two or more persons who have a relationship (such as husband and wife, or parents and children), counselors clarify at the outset which person or persons are clients and the nature of the relationships they will have with each involved person. If it becomes apparent that counselors may be called upon to perform potentially conflicting roles, they clarify, adjust, or withdraw from roles appropriately. (See B.2. and B.4.d.)

### A.9. GROUP WORK

a. *Screening.* Counselors screen prospective group counseling/therapy participants. To the extent possible, counselors select members whose needs and goals are compatible with goals of the group, who will not impede the group process, and whose well-being will not be jeopardized by the group experience.

b. *Protecting Clients.* In a group setting, counselors take reasonable precautions to protect clients from physical or psychological trauma.

A.10. FEES AND BARTERING (See D.3.a. and D.3.b.)

a. *Advance Understanding.* Counselors clearly explain to clients, prior to entering the counseling relationship, all financial arrangements related to professional services, including the use of collection agencies or legal measures for nonpayment. (See A.11.c.)

b. *Establishing Fees.* In establishing fees for professional counseling services, counselors consider the financial status of clients and locality. In the event that the established fee structure is inappropriate for a client, assistance is provided in attempting to find comparable services of acceptable cost. (See A.10.d., D.3.a., and D.3.b.)

c. *Bartering Discouraged.* Counselors ordinarily refrain from accepting goods or services from clients in return for counseling services because such arrangements create inherent potential for conflicts, exploitation, and distortion of the professional relationship. Counselors may participate in bartering only if the relationship is not exploitive, if the client requests it, if a clear written contract is established, and if such arrangements are an accepted practice among professionals in the community. (See A.6.a.)

d. *Pro Bono Service.* Counselors contribute to society by devoting a portion of their professional activity to services for which there is little or no financial return (pro bono).

## A.11. TERMINATION AND REFERRAL

a. *Abandonment Prohibited.* Counselors do not abandon or neglect clients in counseling. Counselors assist in making appropriate arrangements for the continuation of treatment, when necessary, during interruptions, such as vacations, and following termination.

b. *Inability to Assist Clients.* If counselors determine an inability to be of professional assistance to clients, they avoid entering or immediately terminate a counseling relationship. Counselors are knowledgeable about referral resources and suggest appropriate alternatives. If clients decline the suggested referral, counselors should discontinue the relationship.

c. *Appropriate Termination.* Counselors terminate a counseling relationship, securing client agreement when possible, when it is reasonably clear that the client is no longer benefiting, when services are no longer required, when counseling no longer serves the client's needs or interests, when clients do not pay fees charged, or when agency or institution limits do not allow provision of further counseling services. (See A.10.b. and C.2.g.)

## A.12. COMPUTER TECHNOLOGY

a. *Use of Computers.* When computer applications are used in counseling services, counselors ensure that (1) the client is intellectually, emotionally, and physically capable of using the computer application; (2) the computer application is appropriate for the needs of the client; (3) the client understands the purpose and operation of the computer applications; and (4) a follow-up of client use of a computer application is provided to correct possible misconceptions, discover inappropriate use, and assess subsequent needs.

b. *Explanation of Limitations.* Counselors ensure that clients are provided information as a part of the counseling relationship that adequately explains the limitations of computer technology.

c. *Access to Computer Applications.* Counselors provide for equal access to computer applications in counseling services. (See A.2.a.)

## Section B: Confidentiality

### B.1. RIGHT TO PRIVACY

a. *Respect for Privacy.* Counselors respect their clients' right to privacy and avoid illegal and unwarranted disclosures of confidential information. (See A.3.a. and B.6.a.)

b. *Client Waiver.* The right to privacy may be waived by the client or his or her legally recognized representative.

c. *Exceptions.* The general requirement that counselors keep information confidential does not apply when disclosure is required to prevent clear and imminent danger to the client or others or when legal requirements demand that confidential information be revealed. Counselors consult with other professionals when in doubt as to the validity of an exception.

d. *Contagious, Fatal Diseases.* A counselor who receives information confirming that a client has a disease commonly known to be both communicable and fatal is justified in disclosing information to an identifiable third party, who by his or her relationship with the client is at a high risk of contracting the disease. Prior to making a disclosure the counselor should ascertain that the client has not already informed the third party about his or her disease and that the client is not intending to inform the third party in the immediate future. (See B.1.c. and B.1.f.)

e. *Court Ordered Disclosure.* When court ordered to release confidential information without a client's permission, counselors request to the court that the disclosure not be required due to potential harm to the client or counseling relationship. (See B.1.c.)

f. *Minimal Disclosure.* When circumstances require the disclosure of confidential information, only essential information is revealed. To the extent possible, clients are informed before confidential information is disclosed.

g. *Explanation of Limitations.* When counseling is initiated and throughout the counseling process as necessary, counselors inform clients of the limitations of confidentiality and identify foreseeable situations in which confidentiality must be breached. (See G.2.a.)

h. *Subordinates.* Counselors make every effort to ensure that privacy and confidentiality of clients are maintained by subordinates including employees, supervisees, clerical assistants, and volunteers. (See B.1.a.)

i. *Treatment Teams.* If client treatment will involve a continued review by a treatment team, the client will be informed of the team's existence and composition.

### B.2. GROUPS AND FAMILIES

a. *Group Work.* In group work, counselors clearly define confidentiality and the parameters for the specific group being entered, explain its importance, and discuss the difficulties related to confidentiality involved in group work. The fact that confidentiality cannot be guaranteed is clearly communicated to group members.

b. *Family Counseling.* In family counseling, information about one family member cannot be disclosed to another member without permission. Counselors protect the privacy rights of each family member. (See A.8., B.3. and B.4.d.)

## B.3. MINOR OR INCOMPETENT CLIENTS

When counseling clients who are minors or individuals who are unable to give voluntary, informed consent, parents or guardians may be included in the counseling process as appropriate. Counselors act in the best interests of clients and take measures to safeguard confidentiality. (See A.3.c.)

## B.4. RECORDS

a. *Requirement of Records.* Counselors maintain records necessary for rendering professional services to their clients and as required by laws, regulations, or agency or institution procedures.

b. *Confidentiality of Records.* Counselors are responsible for securing the safety and confidentiality of any counseling records they create, maintain, transfer, or destroy whether the records are written, taped, computerized, or stored in any other medium. (See B.1.a.)

c. *Permission to Record or Observe.* Counselors obtain permission from clients prior to electronically recording or observing sessions. (See A.3.a.)

d. *Client Access.* Counselors recognize that counseling records are kept for the benefit of clients and, therefore, provide access to records and copies of records when requested by competent clients unless the records contain information that may be misleading and detrimental to the client. In situations involving multiple clients, access to records is limited to those parts of records that do not include confidential information related to another client. (See A.8., B.1.a., and B.2.b.)

e. *Disclosure or Transfer.* Counselors obtain written permission from clients to disclose or transfer records to legitimate third parties unless exceptions to confidentiality exist as listed in Section B.1. Steps are taken to ensure that receivers of counseling records are sensitive to their confidential nature.

## B.5. RESEARCH AND TRAINING

a. *Data Disguise Required.* Use of data derived from counseling relationships for purposes of training, research, or publication is confined to content that is disguised to ensure the anonymity of the individuals involved. (See B.1.g. and G.3.d.)

b. *Agreement for Identification.* Identification of a client in a presentation or publication is permissible only when the client has reviewed the material and has agreed to its presentation or publication. (See G.3.d.)

## B.6. CONSULTATION

a. *Respect for Privacy.* Information obtained in a consulting relationship is discussed for professional purposes only with persons clearly concerned with the case. Written and oral reports present data germane to the purposes of the consultation, and every effort is made to protect client identity and avoid undue invasion of privacy.

b. *Cooperating Agencies.* Before sharing information, counselors make efforts to ensure that there are defined policies in other agencies serving the counselor's clients that effectively protect the confidentiality of information.

## Section C: Professional Responsibility

## C.1. STANDARDS KNOWLEDGE

Counselors have a responsibility to read, understand, and follow the *Code of Ethics* and the *Standards of Practice.*

## C.2. PROFESSIONAL COMPETENCE

a. *Boundaries of Competence.* Counselors practice only within the boundaries of their competence, based on their education, training, supervised experience, state and national professional credentials, and appropriate professional experience. Counselors will demonstrate a commitment to gain knowledge, personal awareness, sensitivity, and skills pertinent to working with a diverse client population.

b. *New Specialty Areas of Practice.* Counselors practice in specialty areas new to them only after appropriate education, training, and supervised experience. While developing skills in new specialty areas, counselors take steps to ensure the competence of their work and to protect others from possible harm.

c. *Qualified for Employment.* Counselors accept employment only for positions for which they are qualified by education, training, supervised experience, state and national professional credentials, and appropriate professional experience. Counselors hire for professional counseling positions only individuals who are qualified and competent.

d. *Monitor Effectiveness.* Counselors continually monitor their effectiveness as professionals and take steps to improve when necessary. Counselors in private practice take reasonable steps to seek out peer supervision to evaluate their efficacy as counselors.

e. *Ethical Issues Consultation.* Counselors take reasonable steps to consult with other counselors or related professionals when they have questions regarding their ethical obligations or professional practice. (See H.1.)

f. *Continuing Education.* Counselors recognize the need for continuing education to maintain a reasonable level of awareness of current scientific and professional information in their fields of activity. They take steps to maintain competence in the skills they use, are open to new procedures, and keep current with the diverse and/ or special populations with whom they work.

g. *Impairment.* Counselors refrain from offering or accepting professional services when their physical, mental, or emotional problems are likely to harm a client or others. They are alert to the signs of impairment, seek assistance for problems, and, if necessary, limit, suspend, or terminate their professional responsibilities. (See A.11.c.)

## C.3. ADVERTISING AND SOLICITING CLIENTS

a. *Accurate Advertising.* There are no restrictions on advertising by counselors except those that can be specifically justified to protect the public from deceptive practices. Counselors advertise or represent their services to the public by identifying their credentials in an accurate manner that is not false, misleading, deceptive, or fraudulent. Counselors may only advertise the highest degree earned which is in counseling or a closely related field from a college or university that was accredited when the degree was awarded by one of the regional accrediting bodies recognized by the Council on Postsecondary Accreditation.

b. *Testimonials.* Counselors who use testimonials do not solicit them from clients or other persons who, because of their particular circumstances, may be vulnerable to undue influence.

c. *Statements by Others.* Counselors make reasonable efforts to ensure that statements made by others about them or the profession of counseling are accurate.

d. *Recruiting Through Employment.* Counselors do not use their places of employment or institutional affiliation to recruit or gain clients, supervisees, or consultees for their private practices. (See C.5.e.)

e. *Products and Training Advertisements.* Counselors who develop products related to their profession or conduct workshops or training events ensure that the advertisements concerning these products or events are accurate and disclose adequate information for consumers to make informed choices.

f. *Promoting to Those Served.* Counselors do not use counseling, teaching, training, or supervisory relationships to promote their products or training events in a manner that is deceptive or would exert undue influence on individuals who may be vulnerable. Counselors may adopt textbooks they have authored for instruction purposes.

g. *Professional Association Involvement.* Counselors actively participate in local, state, and national associations that foster the development and improvement of counseling.

## C.4. CREDENTIALS

a. *Credentials Claimed.* Counselors claim or imply only professional credentials possessed and are responsible for correcting any known misrepresentations of their credentials by others. Professional credentials include graduate degrees in counseling or closely related mental health fields, accreditation of graduate programs, national voluntary certifications, government-issued certifications or licenses, ACA professional membership, or any other credential that might indicate to the public specialized knowledge or expertise in counseling.

b. *ACA Professional Membership.* ACA professional members may announce to the public their membership status. Regular members may not announce their ACA membership in a manner that might imply they are credentialed counselors.

c. *Credential Guidelines.* Counselors follow the guidelines for use of credentials that have been established by the entities that issue the credentials.

d. *Misrepresentation of Credentials.* Counselors do not attribute more to their credentials than the credentials represent and do not imply that other counselors are not qualified because they do not possess certain credentials.

e. *Doctoral Degrees from Other Fields.* Counselors who hold a master's degree in counseling or a closely related mental health field but hold a doctoral degree from other than counseling or a closely related field do not use the title "Dr." in their practices and do not announce to the public in relation to their practice or status as a counselor that they hold a doctorate.

## C.5. PUBLIC RESPONSIBILITY

a. *Nondiscrimination.* Counselors do not discriminate against clients, students, or supervisees in a manner that has a negative impact based on their age, color, culture, disability, ethnic group, gender, race, religion, sexual orientation, or socioeconomic status, or for any other reason (See A.2.a.)

b. *Sexual Harassment.* Counselors do not engage in sexual harassment. Sexual harassment is defined as sexual solicitation, physical advances, or verbal or nonverbal conduct that is sexual in nature, that occurs in connection with professional activities or roles, and that either (1) is unwelcome, is offensive, or creates a hostile workplace environment, and counselors know or are told this; or (2) is sufficiently severe or intense to be perceived as harassment by a reasonable person in the context. Sexual harassment can consist of a single intense or severe act or multiple persistent or pervasive acts.

c. *Reports to Third Parties.* Counselors are accurate, honest, and unbiased in reporting their professional activities and judgments to appropriate third parties including courts, health insurance companies, those who are the recipients of evaluation reports, and others. (See B.1.g.)

d. *Media Presentations.* When counselors provide advice or comment by means of public lectures, demonstrations, radio or television programs, prerecorded tapes, printed articles, mailed material, or other media, they take reasonable precautions to ensure that (1) the statements are based on appropriate professional counseling literature and practice; (2) the statements are otherwise consistent with the *Code of Ethics* and the *Standards of Practice*; and (3) the recipients of the information are not encouraged to infer that a professional counseling relationship has been established. (See C.6.b.)

e. *Unjustified Gains.* Counselors do not use their professional positions to seek or receive unjustified personal gains, sexual favors, unfair advantage, or unearned goods or services. (See C.3.d.)

## C.6. RESPONSIBILITY TO OTHER PROFESSIONALS

a. *Different Approaches.* Counselors are respectful of approaches to professional counseling that differ from their own. Counselors know and take into account the traditions and practices of other professional groups with which they work.

b. *Personal Public Statements.* When making personal statements in a public context, counselors clarify that they are speaking from their personal perspectives and that they are not speaking on behalf of all counselors or the profession. (See C.5.d.)

c. *Clients Served by Others.* When counselors learn that their clients are in a professional relationship with another mental health professional, they request release from clients to inform the other professionals and strive to establish positive and collaborative professional relationships. (See A.4.)

## Section D: Relationships with Other Professionals

### D.1. RELATIONSHIPS WITH EMPLOYERS AND EMPLOYEES

a. *Role Definition.* Counselors define and describe for their employers and employees the parameters and levels of their professional roles.

b. *Agreements.* Counselors establish working agreements with supervisors, colleagues, and subordinates regarding counseling or clinical relationships, confidentiality, adherence to professional standards, distinction between public and private material, maintenance and dissemination of recorded information, workload, and accountability. Working agreements in each instance are specified and made known to those concerned.

c. *Negative Conditions.* Counselors alert their employers to conditions that may be potentially disruptive or damaging to the counselor's professional responsibilities or that may limit their effectiveness.

d. *Evaluation.* Counselors submit regularly to professional review and evaluation by their supervisor or the appropriate representative of the employer.

e. *In-Service.* Counselors are responsible for in-service development of self and staff.

f. *Goals.* Counselors inform their staff of goals and programs.

g. *Practices.* Counselors provide personal and agency practices that respect and enhance the rights and welfare of each employee and recipient of agency services. Counselors strive to maintain the highest levels of professional services.

h. *Personnel Selection and Assignment.* Counselors select competent staff and assign responsibilities compatible with their skills and experiences.

i. *Discrimination.* Counselors, as either employers or employees, do not engage in or condone practices that are inhumane, illegal, or unjustifiable (such as considerations based on age, color, culture, disability, ethnic group, gender, race, religion, sexual orientation, or socioeconomic status) in hiring, promotion, or training. (See A.2.a. and C.5.b.)

j. *Professional Conduct.* Counselors have a responsibility both to clients and to the agency or institution within which services are performed to maintain high standards of professional conduct.

k. *Exploitive Relationships.* Counselors do not engage in exploitive relationships with individuals over whom they have supervisory, evaluative, or instructional control or authority.

l. *Employer Policies.* The acceptance of employment in an agency or institution implies that counselors are in agreement with its general policies and principles. Counselors strive to reach agreement with employers as to acceptable standards of conduct that allow for changes in institutional policy conducive to the growth and development of clients.

## D.2. CONSULTATION (See B.6.)

a. *Consultation as an Option.* Counselors may choose to consult with any other professionally competent persons about their clients. In choosing consultants, counselors avoid placing the consultant in a conflict of interest situation that would preclude the consultant being a proper party to the counselor's efforts to help the client. Should counselors be engaged in a work setting that compromises this consultation standard, they consult with other professionals whenever possible to consider justifiable alternatives.

b. *Consultant Competency.* Counselors are reasonably certain that they have or the organization represented has the necessary competencies and resources for giving the kind of consulting services needed and that appropriate referral resources are available.

c. *Understanding with Clients.* When providing consultation, counselors attempt to develop with their clients a clear understanding of problem definition, goals for change, and predicted consequences of interventions selected.

d. *Consultant Goals.* The consulting relationship is one in which client adaptability and growth toward self-direction are consistently encouraged and cultivated. (See A.1.b.)

## D.3. FEES FOR REFERRAL

a. *Accepting Fees from Agency Clients.* Counselors refuse a private fee or other remuneration for rendering services to persons who are entitled to such services through the counselor's employing agency or institution. The policies of a particular agency may make explicit provisions for agency clients to receive counseling services from members of its staff in private practice. In such instances, the clients must be informed of other options open to them should they seek private counseling services. (See A.10.a., A.11.b., and C.3.d.)

b. *Referral Fees.* Counselors do not accept a referral fee from other professionals.

## D.4. SUBCONTRACTOR ARRANGEMENTS

When counselors work as subcontractors for counseling services for a third party, they have a duty to inform clients of the limitations of confidentiality that the organization may place on counselors in providing counseling services to clients. The limits of such confidentiality ordinarily are discussed as part of the intake session. (See B.1.e. and B.1.f.)

# Section E: Evaluation, Assessment, and Interpretation

## E.1. GENERAL

a. *Appraisal Techniques.* The primary purpose of educational and psychological assessment is to provide measures that are objective and interpretable in either comparative or absolute terms. Counselors recognize the need to interpret the statements in this section as applying to the whole range of appraisal techniques including test and nontest data.

b. *Client Welfare.* Counselors promote the welfare and best interests of the client in the development, publication, and utilization of educational and psychological assessment techniques. They do not misuse assessment results and interpretations and take reasonable steps to prevent others from misusing the information these techniques provide. They respect the client's right to know the results, the interpretations made, and the basis for their conclusions and recommendations.

## E.2. COMPETENCE TO USE AND INTERPRET TESTS

a. *Limits of Competence.* Counselors recognize the limits of their competence and perform only those testing and assessment services for which they have been trained. They are familiar with reliability, validity, related standardization, error of measurement, and proper application of any technique utilized. Counselors using computer-based test interpretations are trained in the construct being measured and the specific instrument being used prior to using this type of computer application. Counselors take reasonable measures to ensure the proper use of psychological assessment techniques by persons under their supervision.

b. *Appropriate Use.* Counselors are responsible for the appropriate application, scoring, interpretation, and use of assessment instruments whether they score and interpret such tests themselves or use computerized or other services.

c. *Decisions Based on Results.* Counselors responsible for decisions involving individuals or policies that are based on assessment results have a thorough understanding of educational and psychological measurement, including validation criteria, test research, and guidelines for test development and use.

d. *Accurate Information.* Counselors provide accurate information and avoid false claims or misconceptions when making statements about assessment instruments or techniques. Special efforts are made to avoid unwarranted connotations of such terms as IQ and grade equivalent scores. (See C.5.c.)

## E.3. INFORMED CONSENT

a. *Explanation to Clients.* Prior to assessment, counselors explain the nature and purposes of assessment and the specific use of results in language the client (or other legally authorized person on behalf of the client) can understand unless an explicit exception to this right has been agreed upon in advance. Regardless of whether scoring and interpretation are completed by counselors, by assistants, or by computer or other outside services, counselors take reasonable steps to ensure that appropriate explanations are given to the client.

b. *Recipients of Results.* The examinee's welfare, explicit understanding, and prior agreement determine the recipients of test results. Counselors include accurate and appropriate interpretations with any release of individual or group test results. (See B.1.a. and C.5.c.)

## E.4. RELEASE OF INFORMATION TO COMPETENT PROFESSIONALS

a. *Misuse of Results.* Counselors do not misuse assessment results, including test results, and interpretations, and take reasonable steps to prevent the misuse of such by others. (See C.5.c.)

b. *Release of Raw Data.* Counselors ordinarily release data (e.g., protocols, counseling or interview notes, or questionnaires) in which the client is identified only with the consent of the client or the client's legal representative. Such data are usually released only to persons recognized by counselors as competent to interpret the data. (See B.1.a.)

## E.5. PROPER DIAGNOSIS OF MENTAL DISORDERS

a. *Proper Diagnosis.* Counselors take special care to provide proper diagnosis of mental disorders. Assessment techniques (including personal interview) used to determine client care (e.g., locus of treatment, type of treatment, or recommended follow-up) are carefully selected and appropriately used. (See A.3.a. and C.5.c)

b. *Cultural Sensitivity.* Counselors recognize that culture affects the manner in which clients' problems are defined. Clients' socioeconomic and cultural experience is considered when diagnosing mental disorders.

## E.6. TEST SELECTION

a. *Appropriateness of Instruments.* Counselors carefully consider the validity, reliability, psychometric limitations, and appropriateness of instruments when selecting tests for use in a given situation or with a particular client.

b. *Culturally Diverse Populations.* Counselors are cautious when selecting tests for culturally diverse populations to avoid inappropriateness of testing that may be outside of socialized behavioral or cognitive patterns.

## E.7. CONDITIONS OF TEST ADMINISTRATION

a. *Administration Conditions.* Counselors administer tests under the same conditions that were established in their standardization. When tests are not administered under standard conditions or when unusual behavior or irregularities occur during the testing session, those conditions are noted in interpretation, and the results may be designated as invalid or of questionable validity.

b. *Computer Administration.* Counselors are responsible for ensuring that administration programs function properly to provide clients with accurate results when a computer or other electronic methods are used for test administration. (See A.12.b.)

c. *Unsupervised Test-Taking*. Counselors do not permit unsupervised or inadequately supervised use of tests or assessments unless the tests or assessments are designed, intended, and validated for self-administration and/or scoring.

d. *Disclosure of Favorable Conditions*. Prior to test administration, conditions that produce most favorable test results are made known to the examinee.

## E.8. DIVERSITY IN TESTING

Counselors are cautious in using assessment techniques, making evaluations, and interpreting the performance of populations not represented in the norm group on which an instrument was standardized. They recognize the effects of age, color, culture, disability, ethnic group, gender, race, religion, sexual orientation, and socioeconomic status on test administration and interpretation and place test results in proper perspective with other relevant factors. (See A.2.a.)

## E.9. TEST SCORING AND INTERPRETATION

a. *Reporting Reservations*. In reporting assessment results, counselors indicate any reservations that exist regarding validity or reliability because of the circumstances of the assessment or the inappropriateness of the norms for the person tested.

b. *Research Instruments*. Counselors exercise caution when interpreting the results of research instruments possessing insufficient technical data to support respondent results. The specific purposes for the use of such instruments are stated explicitly to the examinee.

c. *Testing Services*. Counselors who provide test scoring and test interpretation services to support the assessment process confirm the validity of such interpretations. They accurately describe the purpose, norms, validity, reliability, and applications of the procedures and any special qualifications applicable to their use. The public offering of an automated test interpretations service is considered a professional-to-professional consultation. The formal responsibility of the consultant is to the consultee, but the ultimate and overriding responsibility is to the client.

## E.10. TEST SECURITY

Counselors maintain the integrity and security of tests and other assessment techniques consistent with legal and contractual obligations. Counselors do not appropriate, reproduce, or modify published tests or parts thereof without acknowledgment and permission from the publisher.

## E.11. OBSOLETE TESTS AND OUTDATED TEST RESULTS

Counselors do not use data or test results that are obsolete or outdated for the current purpose. Counselors make every effort to prevent the misuse of obsolete measures and test data by others.

## E.12. TEST CONSTRUCTION

Counselors use established scientific procedures, relevant standards, and current professional knowledge for test design in the development, publication, and utilization of educational and psychological assessment techniques.

## Section F: Teaching, Training, and Supervision

### F.1. COUNSELOR EDUCATORS AND TRAINERS

a. *Educators as Teachers and Practitioners.* Counselors who are responsible for developing, implementing, and supervising educational programs are skilled as teachers and practitioners. They are knowledgeable regarding the ethical, legal, and regulatory aspects of the profession, are skilled in applying that knowledge, and make students and supervisees aware of their responsibilities. Counselors conduct counselor education and training programs in an ethical manner and serve as role models for professional behavior. Counselor educators should make an effort to infuse material related to human diversity into all courses and/or workshops that are designed to promote the development of professional counselors.

b. *Relationship Boundaries with Students and Supervisees.* Counselors clearly define and maintain ethical, professional, and social relationship boundaries with their students and supervisees. They are aware of the differential in power that exists and the student's or supervisee's possible incomprehension of that power differential. Counselors explain to students and supervisees the potential for the relationship to become exploitive.

c. *Sexual Relationships.* Counselors do not engage in sexual relationships with students or supervisees and do not subject them to sexual harassment. (See A.6. and C.5.b.)

d. *Contributions to Research.* Counselors give credit to students or supervisees for their contributions to research and scholarly projects. Credit is given through co-authorship, acknowledgment, footnote statement, or other appropriate means in accordance with such contributions. (See G.4.b. and G.4.c.)

e. *Close Relatives.* Counselors do not accept close relatives as students or supervisees.

f. *Supervision Preparation.* Counselors who offer clinical supervision services are adequately prepared in supervision methods and techniques. Counselors who are doctoral students serving as practicum or internship supervisors to master's level students are adequately prepared and supervised by the training program.

g. *Responsibility for Services to Clients.* Counselors who supervise the counseling services of others take reasonable measures to ensure that counseling services provided to clients are professional.

h. *Endorsement.* Counselors do not endorse students or supervisees for certification, licensure, employment, or completion of an academic or training program if they believe students or supervisees are not qualified for the endorsement. Counselors take reasonable steps to assist students or supervisees who are not qualified for endorsement to become qualified.

### F.2. COUNSELOR EDUCATION AND TRAINING PROGRAMS

a. *Orientation.* Prior to admission, counselors orient prospective students to the counselor education or training program's expectations including but not limited to the following: (1) the type and level of skill acquisition required for successful completion of the training, (2) subject matter to be covered, (3) basis for evaluation, (4) training components that encourage self-growth or self-disclosure as part of the training process, (5) the type of supervision settings and requirements of the sites for required clinical field experiences, (6) student and supervisee evaluation and dismissal policies and procedures, and (7) up-to-date employment prospects for graduates.

b. *Integration of Study and Practice.* Counselors establish counselor education and training programs that integrate academic study and supervised practice.

c. *Evaluation.* Counselors clearly state to students and supervisees, in advance of training, the levels of competency expected, appraisal methods, and timing of evaluations for both didactic and experiential components. Counselors provide students and supervisees with periodic performance appraisal and evaluation feedback throughout the training program.

d. *Teaching Ethics.* Counselors make students and supervisees aware of the ethical responsibilities and standards of the profession and the students' and superisees' ethical responsibilities to the profession. (See C.1. and F.3.e.)

e. *Peer Relationships.* When students or supervisees are assigned to lead counseling groups or provide clinical supervision for their peers, counselors take steps to ensure that students and supervisees placed in these roles do not have personal or adverse relationships with peers and that they understand they have the same ethical obligations as counselor educators, trainers, and supervisors. Counselors make every effort to ensure that the rights of peers are not compromised when students or supervisees are assigned to lead counseling groups or provide clinical supervision.

f. *Varied Theoretical Positions.* Counselors present varied theoretical positions so that students and supervisees may make comparisons and have opportunities to develop their own positions. Counselors provide information concerning the scientific basis of professional practice. (See C.6.a.)

g. *Field Placements.* Counselors develop clear policies within their training program regarding field placement and other clinical experiences. Counselors provide clearly stated roles and responsibilities for the student or supervisee, the site supervisor, and the program supervisor. They confirm that site supervisors are qualified to provide supervision and are informed of their professional and ethical responsibilities in this role.

h. *Dual Relationships as Supervisors.* Counselors avoid dual relationships, such as performing the role of site supervisor and training program supervisor in the student's or supervisee's training program. Counselors do not accept any form of professional services, fees, commissions, reimbursement, or remuneration from a site for student or supervisee placement.

i. *Diversity in Programs.* Counselors are responsive to their institution's and program's recruitment and retention needs for training program administrators, faculty, and students with diverse backgrounds and special needs. (See A.2.a.)

## F.3. STUDENTS AND SUPERVISEES

a. *Limitations.* Counselors, through ongoing evaluation and appraisal, are aware of the academic and personal limitations of students and supervisees that might impede performance. Counselors assist students and supervisees in securing remedial assistance when needed and dismiss from the training program supervisees who are unable to provide competent service due to academic or personal limitations. Counselors seek professional consultation and document their decision to dismiss or refer students or supervisees for assistance. Counselors assure that students and supervisees have recourse to address decisions made, to require them to seek assistance, or to dismiss them.

b. *Self-Growth Experiences.* Counselors use professional judgment when designing training experiences conducted by the counselors themselves that require student and supervisee self-growth or self-disclosure. Safeguards are provided so that students and supervisees are aware of the ramifications their self-disclosure may have on counselors whose primary role as teacher, trainer, or supervisor requires acting on ethical obligations to the profession. Evaluative components of experiential training experiences explicitly delineate predetermined academic standards that are separate and not dependent on the student's level of self-disclosure. (See A.6.)

c. *Counseling for Students and Supervisees.* If students or supervisees request counseling, supervisors or counselor educators provide them with acceptable referrals. Supervisors or counselor educators do not serve as counselor to students or supervisees over whom they hold administrative, teaching, or evaluative roles unless this is a brief role associated with a training experience. (See A.6.b.)

d. *Clients of Students and Supervisees.* Counselors make every effort to ensure that the clients at field placements are aware of the services rendered and the qualifications of the students and supervisees rendering those services. Clients receive professional disclosure information and are informed of the limits of confidentiality. Client permission is obtained in order for the students and supervisees to use any information concerning the counseling relationship in the training process. (See B.1.e.)

e. *Standards for Students and Supervisees.* Students and supervisees preparing to become counselors adhere to the *Code of Ethics* and the *Standards of Practice.* Students and supervisees have the same obligations to clients as those required of counselors. (See H.1.)

## Section G: Research and Publication

G.1. RESEARCH RESPONSIBILITIES

a. *Use of Human Subjects.* Counselors plan, design, conduct, and report research in a manner consistent with pertinent ethical principles, federal and state laws, host institutional regulations, and scientific standards governing research with human subjects. Counselors design and conduct research that reflects cultural sensitivity appropriateness.

b. *Deviation from Standard Practices.* Counselors seek consultation and observe stringent safeguards to protect the rights of research participants when a research problem suggests a deviation from standard acceptable practices. (See B.6.)

c. *Precautions to Avoid Injury.* Counselors who conduct research with human subjects are responsible for the subjects' welfare throughout the experiment and take reasonable precautions to avoid causing injurious psychological, physical, or social effects to their subjects.

d. *Principal Researcher Responsibility.* The ultimate responsibility for ethical research practice lies with the principle researcher. All others involved in the research activities share ethical obligations and full responsibility for their own actions.

e. *Minimal Interference.* Counselors take reasonable precautions to avoid causing disruptions in subjects' lives due to participation in research.

f. *Diversity.* Counselors are sensitive to diversity and research issues with special populations. They seek consultation when appropriate. (See A.2.a. and B.6.)

G.2. INFORMED CONSENT

a. *Topics Disclosed.* In obtaining informed consent for research, counselors use language that is understandable to research participants and that (1) accurately explains the purpose and procedures to be followed; (2) identifies any procedures that are experimental or relatively untried; (3) describes the attendant discomforts and risks; (4) describes the benefits or changes in individuals or organizations that might be reasonably expected; (5) discloses appropriate alternative procedures that would be advantageous for subjects; (6) offers to answer any inquiries concerning the procedures; (7) describes any limitations on confidentiality; and (8) instructs that subjects are free to withdraw their consent and to discontinue participation in the project at any time. (See B.1.f.)

b. *Deception.* Counselors do not conduct research involving deception unless alternative procedures are not feasible and the prospective value of the research justifies the deception. When the methodological requirements of a study necessitate concealment or deception, the investigator is required to explain clearly the reasons for this action as soon as possible.

c. *Voluntary Participation.* Participation in research is typically voluntary and without any penalty for refusal to participate. Involuntary participation is appropriate only when it can be demonstrated that participation will have no harmful effects on subjects and is essential to the investigation.

d. *Confidentiality of Information.* Information obtained about research participants during the course of an investigation is confidential. When the possibility exists that others may obtain access to such information, ethical research practice requires that the possibility, together with the plans for protecting confidentiality, be explained to participants as a part of the procedure for obtaining informed consent. (See B.1.e.)

e. *Persons Incapable of Giving Informed Consent.* When a person is incapable of giving informed consent, counselors provide an appropriate explanation, obtain agreement for participation and obtain appropriate consent from a legally authorized person.

f. *Commitments to Participants.* Counselors take reasonable measures to honor all commitments to research participants.

g. *Explanations After Data Collection.* After data are collected, counselors provide participants with full clarification of the nature of the study to remove any misconceptions. Where scientific or human values justify delaying or withholding information, counselors take reasonable measures to avoid causing harm.

h. *Agreements to Cooperate.* Counselors who agree to cooperate with another individual in research or publication incur an obligation to cooperate as promised in terms of punctuality of performance and with regard to the completeness and accuracy of the information required.

i. *Informed Consent for Sponsors.* In the pursuit of research counselors give sponsors, institutions, and publication channels the same respect and opportunity for giving informed consent that they accord to individual research participants. Counselors are aware of their obligation to future research workers and ensure that host institutions are given feedback information and proper acknowledgment.

## G.3. REPORTING RESULTS

a. *Information Affecting Outcome.* When reporting research results counselors explicitly mention all variables and conditions known to the investigator that may have affected the outcome of a study or the interpretation of data.

b. *Accurate Results.* Counselors plan, conduct, and report research accurately and in a manner that minimizes the possibility that results will be misleading. They provide thorough discussions of the limitations of their data and alternative hypotheses.

Counselors do not engage in fraudulent research, distort data, misrepresent data, or deliberately bias their results.

c. *Obligation to Report Unfavorable Results.* Counselors communicate to other counselors the results of any research judged to be of professional value. Results that reflect unfavorably on institutions, programs, services, prevailing opinions, or vested interests are not withheld.

d. *Identity of Subjects.* Counselors who supply data, aid in the research of another person, report research results, or make original data available take due care to disguise the identity of respective subjects in the absence of specific authorization from the subjects to do otherwise. (See B.1.g. and B.5.a.)

e. *Replication Studies.* Counselors are obligated to make available sufficient original research data to qualified professionals who may wish to replicate the study.

## G.4. PUBLICATION

a. *Recognition of Others.* When conducting and reporting research, counselors are familiar with and give recognition to previous work on the topic, observe copyright laws, and give full credit to those to whom credit is due. (See F.1.d. and G.4.c.)

b. *Contributors.* Counselors give credit through joint authorship, acknowledgment, footnote statements, or other appropriate means to those who have contributed significantly to research or concept development in accordance with such contributions. The principal contributor is listed first, and minor technical or professional contributions are acknowledged in notes or introductory statements.

c. *Student Research.* For an article that is substantially based on a student's dissertation or thesis, the student is listed as the principal author. (See F.1.d. and G.4.a.)

d. *Duplicate Submission.* Counselors submit manuscripts for consideration to only one journal at a time. Manuscripts that are published in whole or in substantial part in another journal or published work are not submitted for publication without acknowledgment and permission from the previous publication.

e. *Professional Review.* Counselors who review material submitted for publication, research, or other scholarly purposes respect the confidentiality and proprietary rights of those who submitted it.

## Section H: Resolving Ethical Issues

### H.1. KNOWLEDGE OF STANDARDS

Counselors are familiar with the *Code of Ethics* and the *Standards of Practice* and other applicable ethics codes from other professional organizations of which they are member or from certification and licensure bodies. Lack of knowledge or misunderstanding of an ethical responsibility is not a defense against a charge of unethical conduct. (See F.3.e.)

### H.2. SUSPECTED VIOLATIONS

a. *Ethical Behavior Expected.* Counselors expect professional associates to adhere to the *Code of Ethics.* When counselors possess reasonable cause that raises doubts as to whether a counselor is acting in an ethical manner, they take appropriate action. (See H.2.d. and H.2.e.)

b. *Consultation.* When uncertain as to whether a particular situation or course of action may be in violation of the *Code of Ethics,* counselors consult with other counselors who are knowledgeable about ethics, with colleagues, or with appropriate authorities.

c. *Organization Conflicts*. If the demands of an organization with which counselors are affiliated pose a conflict with the *Code of Ethics*, counselors specify the nature of such conflicts and express to their supervisors or other responsible officials their commitment to the *Code of Ethics*. When possible, counselors work toward change within the organization to allow full adherence to the *Code of Ethics*.

d. *Informal Resolution*. When counselors have reasonable cause to believe that another counselor is violating an ethical standard, they attempt to first resolve the issue informally with the other counselor if feasible providing that such action does not violate confidentiality rights that may be involved.

e. *Reporting Suspected Violations*. When an informal resolution is not appropriate or feasible, counselors, upon reasonable cause, take action, such as reporting the suspected ethical violation to state or national ethics committees, unless this action conflicts with confidentiality rights that cannot be resolved.

f. *Unwarranted Complaints*. Counselors do not initiate, participate in, or encourage the filing of ethics complaints that are unwarranted or intended to harm a counselor rather than to protect clients or the public.

## H.3. COOPERATION WITH ETHICS COMMITTEES
Counselors assist in the process of enforcing the *Code of Ethics*. Counselors cooperate with investigations, proceedings, and requirements of the ACA Ethics Committee or ethics committees of other duly constituted associations or boards having jurisdiction over those charged with a violation. Counselors are familiar with the ACA Policies and Procedures and use them as a reference in assisting the enforcement of the *Code of Ethics*.

# STANDARDS OF PRACTICE

All members of the American Counseling Association (ACA) are required to adhere to the *Standards of Practice* and the *Code of Ethics*. The *Standards of Practice* represent minimal behavioral statements of the *Code of Ethics*. Members should refer to the applicable section of the *Code of Ethics* for further interpretation and amplification of the applicable *Standards of Practice*.

## Section A: The Counseling Relationship

**Standard of Practice One (SP-1): Nondiscrimination**
Counselors respect diversity and must not discriminate against clients because of age, color, culture, disability, ethnic group, gender, race, religion, sexual orientation, marital status, or socioeconomic status. (See A.2.a.)

**Standard of Practice Two (SP-2): Disclosure to Clients**
Counselors must adequately inform clients, preferably in writing, regarding the counseling process and counseling relationship at or before the time it begins and throughout the relationship. (See A.3.a.)

**Standard of Practice Three (SP-3): Dual Relationships**
Counselors must make every effort to avoid dual relationships with clients that could impair their professional judgment or increase the risk of harm to clients. When a dual

relationship cannot be avoided, counselors must take appropriate steps to ensure that judgment is not impaired and that no exploitation occurs. (See A.6.a. and A.6.b.)

### Standard of Practice Four (SP-4): Sexual Intimacies with Clients

Counselors must not engage in any type of sexual intimacies with current clients and must not engage in sexual intimacies with former clients within a minimum of two years after terminating the counseling relationship. Counselors who engage in such relationship after two years following termination have the responsibility to thoroughly examine and document that such relations did not have an exploitative nature.

### Standard of Practice Five (SP-5): Protecting Clients During Group Work

Counselors must take steps to protect clients from physical or psychological trauma resulting from interactions during group work. (See A.9.b.)

### Standard of Practice Six (SP-6): Advance Understanding of Fees

Counselors must explain to clients, prior to their entering the counseling relationship, financial arrangements related to professional services. (See A.10.a-d. and A.11.c.)

### Standard of Practice Seven (SP-7): Termination

Counselors must assist in making appropriate arrangements for the continuation of treatment of clients, when necessary, following termination of counseling relationships. (See A.11.a.)

### Standard of Practice Eight (SP-8): Inability to Assist Clients

Counselors must avoid entering or immediately terminate a counseling relationship if it is determined that they are unable to be of professional assistance to a client. The counselor may assist in making an appropriate referral for the client. (See A.11.b.)

## Section B: Confidentiality

### Standard of Practice Nine (SP-9): Confidentiality Requirement

Counselors must keep information related to counseling services confidential unless disclosure is in the best interest of clients, is required for the welfare of others, or is required by law. When disclosure is required, only information that is essential is revealed, and the client is informed of such disclosure (See B.1.a-f.)

### Standard of Practice Ten (SP-10):
### Confidentiality Requirements for Subordinates

Counselors must take measures to ensure that privacy and confidentiality of clients are maintained by subordinates. (See B.1.h.)

### Standard of Practice Eleven (SP-11): Confidentiality in Group Work

Counselors must clearly communicate to group members that confidentiality cannot be guaranteed in group work. (See B.2.a.)

### Standard of Practice Twelve (SP-12): Confidentiality in Family Counseling

Counselors must not disclose information about one family member in counseling to another family member without prior consent. (See B.2.b.)

### Standard of Practice Thirteen (SP-13): Confidentiality of Records

Counselors must maintain appropriate confidentiality in creating, storing, accessing, transferring, and disposing of counseling records. (See B.4.b.)

**Standard of Practice Fourteen (SP-14): Permission to Record or Observe**
Counselors must obtain prior consent from clients in order to electronically record or observe sessions. (See B.4.c.)

**Standard of Practice Fifteen (SP-15): Disclosure or Transfer of Records**
Counselors must obtain client consent to disclose or transfer records to third parties unless exceptions listed in SP-9 exist. (See B.4.e.)

**Standard of Practice Sixteen (SP-16): Data Disguise Required**
Counselors must disguise the identity of the client when using data for training, research, or publication. (See B.5.a.)

## Section C: Professional Responsibility

**Standard of Practice Seventeen (SP-17): Boundaries of Competence**
Counselors must practice only within the boundaries of their competence. (See C.2.a.)

**Standard of Practice Eighteen (SP-18): Continuing Education**
Counselors must engage in continuing education to maintain their professional competence. (See C.2.f.)

**Standard of Practice Nineteen (SP-19): Impairment of Professionals**
Counselors must refrain from offering professional services when their personal problems or conflicts may cause harm to client or others. (See C.2.g.)

**Standard of Practice Twenty (SP-20): Accurate Advertising**
Counselors must accurately represent their credentials and services when advertising. (See C.3.a.)

**Standard of Practice Twenty-one (SP-21): Recruiting Through Employment**
Counselors must not use their place of employment or institutional affiliation to recruit clients for their private practices. (See C.3.d.)

**Standard of Practice Twenty-two (SP-22): Credentials Claimed**
Counselors must claim or imply only professional credentials possessed and must correct any known misrepresentations of their credentials by others. (See C.4.a.)

**Standard of Practice Twenty-three (SP-23): Sexual Harassment**
Counselors must not engage in sexual harassment. (See C.5.b.)

**Standard of Practice Twenty-four (SP-24): Unjustified Gains**
Counselors must not use their professional positions to seek or receive unjustified personal gains, sexual favors, unfair advantage, or unearned goods or services. (See C.5.e.)

**Standard of Practice Twenty-five (SP-25): Clients Served by Others**
With the consent of the client, counselors must inform other mental health professionals serving the same client that a counseling relationship between the counselor and client exists. (See C.6.c.)

**Standard of Practice Twenty-six (SP-26): Negative Employment Conditions**
Counselors must alert their employers to institutional policy or conditions that may be

potentially disruptive or damaging to the counselor's professional responsibilities or that may limit their effectiveness or deny clients' rights. (See D.1.c.)

### Standard of Practice Twenty-seven (SP -27): Personnel Selection and Assignment

Counselors must select competent staff and must assign responsibilities compatible with staff skills and experiences. (See D.1.h.)

### Standard of Practice Twenty-eight (SP-28): Exploitative Relationships with Subordinates

Counselors must not engage in exploitative relationships with individuals over whom they have supervisory, evaluative, or instructional control or authority. (See D.1.k.)

## Section D: Relationship with Other Professionals

### Standard of Practice Twenty-nine (SP-29): Accepting Fees from Agency Clients

Counselors must not accept fees or other remuneration for consultation with persons entitled to such services through the counselor's employing agency or institution. (See D.3.a.)

### Standard of Practice Thirty (SP-30): Referral Fees

Counselors must not accept referral fees. (See D.3.b.)

## Section E: Evaluation, Assessment, and Interpretation

### Standard of Practice Thirty-one (SP-31): Limits of Competence

Counselors must perform only testing and assessment services for which they are competent. Counselors must not allow the use of psychological assessment techniques by unqualified persons under their supervision. (See E.2.a.)

### Standard of Practice Thirty-two (SP-32): Appropriate Use of Assessment Instruments

Counselors must use assessment instruments in the manner for which they were intended. (See E.2.b.)

### Standard of Practice Thirty-three (SP-33): Assessment Explanations to Clients

Counselors must provide explanations to clients prior to assessment about the nature and purposes of assessment and the specific uses of results. (See E.3.a.)

### Standard of Practice Thirty-four (SP-34): Recipients of Test Results

Counselors must ensure that accurate and appropriate interpretations accompany any release of testing and assessment information. (See E.3.b.)

### Standard of Practice Thirty-five (SP-35): Obsolete Tests and Outdated Test Results

Counselors must not base their assessment or intervention decisions or recommendations on data or test results that are obsolete or outdated for the current purpose. (See E.11.)

# Section F: Teaching, Training, and Supervision

**Standard of Practice Thirty-six (SP-36):**
**Sexual Relationships with Students or Supervisees**
Counselors must not engage in sexual relationships with their students and supervisees. (See F.1.c.)

**Standard of Practice Thirty-seven (SP-37):**
**Credit for Contributions to Research**
Counselors must give credit to students or supervisees for their contributions to research and scholarly projects. (See F.1.d.)

**Standard of Practice Thirty-eight (SP-38): Supervision Preparation**
Counselors who offer clinical supervision services must be trained and prepared in supervision methods and techniques. (See F.1.f.)

**Standard of Practice Thirty-nine (SP-39): Evaluation Information**
Counselors must clearly state to students and supervisees, in advance of training, the levels of competency expected, appraisal methods, and timing of evaluations. Counselors must provide students and supervisees with periodic performance appraisal and evaluation feedback throughout the training program. (See F.2.c.)

**Standard of Practice Forty (SP-40): Peer Relationships in Training**
Counselors must make every effort to ensure that the rights of peers are not violated when students and supervisees are assigned to lead counseling groups or provide clinical supervision. (See F.2.e.)

**Standard of Practice Forty-one (SP-41):**
**Limitations of Students and Supervisees**
Counselors must assist students and supervisees in securing remedial assistance, when needed, and must dismiss from the training program students and supervisees who are unable to provide competent service due to academic or personal limitations. (See F.3.a.)

**Standard of Practice Forty-two (SP-42): Self-Growth Experiences**
Counselors who conduct experiences for students or supervisees that include self-growth or self-disclosure must inform participants of counselors' ethical obligations to the profession and must not grade participants based on their nonacademic performance. (See F.3.b.)

**Standard of Practice Forty-three (SP-43):**
**Standards for Students and Supervisees**
Students and supervisees preparing to become counselors must adhere to the *Code of Ethics* and the *Standards of Practice* of counselors. (See F.3.e.)

# Section G: Research and Publication

**Standard of Practice Forty-four (SP-44):**
**Precautions to Avoid Injury in Research**
Counselors must avoid causing physical, social, or psychological harm or injury to subjects in research. (See G.1.c.)

**Standard of Practice Forty-five (SP-45):**
**Confidentiality of Research Information**
Counselors must keep confidential information obtained about research participants. (See G.2.d.)

**Standard of Practice Forty-six (SP-46):**
**Information Affecting Research Outcome**
Counselors must report all variables and conditions known to the investigator that may have affected research data or outcomes. (See G.3.a.)

**Standard of Practice Forty-seven (SP-47): Accurate Research Results**
Counselors must not distort or misrepresent research data, nor fabricate or intentionally bias research results. (See G.3.b.)

**Standard of Practice Forty-eight (SP-48): Publication Contributors**
Counselors must give appropriate credit to those who have contributed to research. (See G.4.a. and G.4.b.)

## Section H: Resolving Ethical Issues

**Standard of Practice Forty-nine (SP-49): Ethical Behavior Expected**
Counselors must take appropriate action when they possess reasonable cause that raises doubts as to whether counselors or other mental health professionals are acting in an ethical manner. (See H.2.a.)

**Standard of Practice Fifty (SP-50): Unwarranted Complaints**
Counselors must not initiate, participate in, or encourage the filing of ethics complaints that are unwarranted or intended to harm a mental health professional rather than to protect clients or the public. (See H.2.f.)

**Standard of Practice Fifty-one (SP-51):**
**Cooperation with Ethics Committees**
Counselors must cooperate with investigations, proceedings, and requirements of the ACA Ethics Committee or ethics committees of other duly constituted associations or boards having jurisdiction over those charged with a violation. (See H.3.)

## REFERENCES

The following documents are available to counselors as resources to guide them in their practices. These resources are not a part of the *Code of Ethics* and the *Standards of Practice*.

AMERICAN ASSOCIATION FOR COUNSELING AND DEVELOP-MENT/ASSOCIATION FOR MEASUREMENT AND EVALUA-TION IN COUNSELING AND DEVELOPMENT. (1989). *The responsibilities of users of standardized tests* (revised). Washington, DC.

AMERICAN COUNSELING ASSO-CIATION. (1988). *American Counseling Association Ethical Standards.* Alexandria, VA.

AMERICAN PSYCHOLOGICAL AS-SOCIATION. (1985). *Standards for educational and psychological testing* (revised). Washington, DC.

AMERICAN REHABILITATION COUNSELING ASSOCIATION, COMMISSION ON REHABILITATION COUNSELOR CERTIFICATION, AND NATIONAL REHABILITATION COUNSELING ASSOCIATION. (1995). *Code of professional ethics for rehabilitation counselors*. Chicago, IL.

AMERICAN SCHOOL COUNSELOR ASSOCIATION. (1992). *Ethical standards for school counselors*. Alexandria, VA.

JOINT COMMITTEE ON TESTING PRACTICES. (1988). *Code of fair testing practices in education*. Washington, DC.

NATIONAL BOARD FOR CERTIFIED COUNSELORS. (1989). *National board for Certified Counselors Code of Ethics*. Alexandria, VA.

PREDIGER, D.J. (Ed.). (1993, March). *Multicultural assessment standards*. Alexandria, VA: Association for Assessment in Counseling.

# APPENDIX F

# State and National Credentialing Boards, Professional Organizations, and Honorary Societies

## STATE CREDENTIALING BOARDS

**ALABAMA**
Board of Examiners in Counseling
P.O. Box 550397
Birmingham, AL 35255
(205) 933-8100
(205) 933-6700 (fax)

**ARIZONA**
Counselor Credentialing Committee
of the Board of Behavioral Examiners
1645 W. Jefferson Ave., 4th Floor
Phoenix, AZ 85007
(602) 542-1882
(602) 542-1830 (fax)

**ARKANSAS**
Board of Examiners in Counseling
Southern Arkansas University
P.O. Box 1396
Magnolia, AR 71753
(501) 235-4314
(501) 234-1842 (fax)
AKThomas@SAUMag.edu(E-MAIL)

**CALIFORNIA**
Board of Behavioral Science
Examiners
400 R Street, Suite 3150
Sacramento, CA 95814-6240
(916) 445-4933

**COLORADO**
Board of Licensed Professional
Counselor Examiners
1560 Broadway, Suite 1340
Denver, CO 80202
(303) 894-7766
(330) 894-7790 (fax)

**CONNECTICUT**
CT Dept. of Public Health
P.O. Box 310308
Hartford, CT 06134
(860) 509-7579

**DELAWARE**
Board of Professional Counselors
of Mental Health
P.O. Box 1401
Cannon Bldg.
Dover, DE 19903
(302) 739-4522
(302) 739-2711 (fax)

**DISTRICT OF COLUMBIA**
DC Board of Professional Counselors
614 H ST., NW, Rm. 108
Washington, DC 20001
(202) 727-5365
(202) 727-4087(fax)

**FLORIDA**
Board of Clinical Social Workers,
Marriage & Family Therapists, &
Mental Health Counselors
Agency for Health Counselors
1940 N. Monroe Street
Tallahassee, FL 32399-0753
(904) 487-2520
(904) 921-5389 (fax)★
Fax documents are not accepted.

**GEORGIA**
Composite Board of Professional
Counselors, Social Workers, and
Marriage & Family Therapists
166 Pryor Street, SW
Atlanta, GA 30303
(404) 656-3933
(404) 651-9532 (fax)

**IDAHO**
Idaho State Counselor Licensure
Board
Bureau of Occupational Licenses
1109 Main Street, Suite 220
Boise, ID 83702-5642
(208) 334-3233
(208) 334-3945 (fax)

**ILLINOIS**
Professional Counselor Licensing
& Disciplinary Board
320 W. Washington St.
Springfield, IL 62786
(217) 785-0872
(217) 782-7645 (fax)

**INDIANA**
Social Worker, MFT's, & Mental
Health Counselor Board
Health Professions Bureau
402 W. Washington ST., Rm. 041
Indianapolis, IN 46204
(317) 232-2960
(317) 233-4236 (fax)

**IOWA**
IA Behavioral Science Board
Lucas Bldg., 4th Fl.
Des Moines, IA 50319
(515) 281-6352
(515) 281-3121 (fax)

**KANSAS**
Behavioral Sciences Regulatory Board
712 S. Kansas Ave.
Topeka, KS 66603
(913) 296-3240
(913) 296-3112 (fax)

**KENTUCKY**
Division of Occupations and
Professions
P.O. Box 456
Frankfort, KY 40602
(502) 564-3296, X226
(502) 564-4818 (fax)

**LOUISIANA**
Licensed Professional Counselors
Board of Examiners
8631 Summa Ave., Ste. A
Baton Rouge, LA 70809
(504) 765-2515
(504) 765-2514 (fax)

## MAINE
Board of Counseling Professionals
State House, Station #35
Augusta, ME 04333
(207) 624-8603
(207) 624-8637 (fax)

## MARYLAND
Board of Examiners of Professional
Counselors
Metro Executive Center, 3rd Floor
4201 Patterson Avenue
Baltimore, MD 21215
(410) 764-4732
(410) 764-5987 (fax)

## MASSACHUSETTS
Board of Allied Mental Health
Services
100 Cambridge Street, 15th Floor
Boston, MA 02202
(617) 727-3080
(617) 727-2197 (fax)

## MICHIGAN
MI Board of Counseling
P.O. Box 30018
Lansing, MI 48909
(517) 335-0918 (applications)
(517) 373-3596 (fax)

## MISSISSIPPI
MS Board of Examiners for LPCs
1101 Robert E. Lee Bldg.
239 N. Lamar St.
Jackson, MS 39201
(601) 359-6630

## MISSOURI
Missouri Committee for Professional
Counselors
P.O. Box 1135
Jefferson City, MO 65102
(573) 751-0018
(573) 751-4176
(573) 526-3489 (fax)

## MONTANA
Board of Social Work Examiners &
Professional Counselors
Arcade Bldg., Lower Level
111 North Jackson
P.O. Box 200513
Helena, MT 59620-0513
(406) 444-4285
(406) 444-1667 (fax)

## NEBRASKA
Board of Examiners in Mental Health
Practice
Prof. & Occup. Licensure Division
301 Centennial Mall South
P.O. Box 95007
Lincoln, NE 68509
(402) 471-2117
(402) 471-3577 (fax)

## NEW HAMPSHIRE
NH Board of Examiners of Psych. &
Mental Health Practice
105 Pleasant Street
Concord, NH 03301
(603) 271-6762

## NEW JERSEY
NJ Professional Counselor Examiner's
Committee
Division of Consumer Affairs
P.O. Box 45033
Newark, NJ 07101
(973) 504-6415
(973) 648-3536 (fax)

## NEW MEXICO
Counselor Therapy & Practice Board
P.O. Box 25101
Santa Fe, NM 87504
(505) 827-7554
(505) 827-7560 (fax)

## NORTH CAROLINA
NC Board of Licensed Professional
Counselors
P.O. Box 21005
Raleigh, NC 27619
(919) 870-1980 (Mon.–Thurs.,
10 A.M.–2 P.M.)
(919) 571-8672 (fax)

## NORTH DAKOTA
ND Board of Counselor Examiners
P.O. Box 2735
Bismarck, ND 58502
(701) 224-8234

## OHIO
Counselor & Social Worker Board
77 South High Street, 16th Floor
Columbus, OH 43266-0340
(614) 466-0912
(614) 728-7790 (fax)

## OKLAHOMA
Licensed Professional Counselors,
Licensed Marital & Family Therapists
1000 NE 10th Street
Oklahoma City, OK 73117-1299
(405) 271-6030
(405) 271-1011 (fax)

## OREGON
Board of Licensed Professional
Counselors and Therapists
3218 Pringle Rd. SE, #160
Salem, OR 97302
(503) 378-5499

## RHODE ISLAND
Board of Mental Health Counselors &
M&F Therapists
Division of Professional Regulation
3 Capitol Hill
Cannon Bldg., Room 104
Providence, RI 02908-5097
(401) 222-2827
(401) 222-1272

## SOUTH CAROLINA
SC Dept. of LLR, Division of POL
Board of Examiners for LPC, AC and
MFT
P.O. Box 11329
Columbia, SC 29211
(803) 734-4243
(803) 734-4284 (fax)

## SOUTH DAKOTA
South Dakota Board of Counselor
Examiners
P.O. Box 1822
Sioux Falls, SD 57101
(605) 331-2927
(605) 331-2043 (fax)

## TENNESSEE
State Board of Professional Counselors
& MFTs
425 5th Ave., N.
Cordell Hull Bldg., 1st Fl.
Nashville, TN 37247-1010
(615) 532-5132
(615) 532-5164 (fax)

## TEXAS
Board of Examiners of Professional
Counselors
1100 W. 49th Street
Austin, TX 78756-3183
(512) 834-6658
(512) 834-6677 (fax)

## UTAH
Division of Occupational and
Professional Licensing
160 E. 300th St.
Salt Lake City, UT 84111
(801) 530-6736

**VERMONT**
LCMHC Advisory Board
Office of Professional Regulation
109 State Street
Montpelier, VT 05609-1106
1-800-439-8683 (in-state)
(802) 828-2390
(802) 828-2496 (fax)

**VIRGINIA**
Board of Professional Counselors &
Marriage and Family Therapists
Dept. of Health Professions
6606 W. Broad Street, 4th Floor
Richmond, VA 23230-1717
(804) 662-9912
(804) 662-9943 (fax)

**WASHINGTON**
Counselor Programs, Dept. of Health
Health Profession Quality Assurance
Division
P.O. Box 47869
Olympia, WA 98504-7869
(360) 664-9098
(360) 586-7775 (fax)

**WEST VIRGINIA**
Board of Examiners in Counseling
WV Graduate College
100 Angus E. Peyton Dr., Rm. 201-D
S. Charleston, WV 25303
(304) 746-2512
1-800-520-3852
(304) 746-1942 (fax)

**WISCONSIN**
Examining Board of Social Workers,
MFT's & Professional Counselors
Dept. of Regulations & Licensing
1400 E. Washington Ave.
P.O. Box 8935
Madison, WI 53708
(608) 267-7212
(608) 267-0644 (fax)

**WYOMING**
Occupational Licensing Director
Mental Health Professions Licensing
Board
2020 Carey Ave., Ste 201
Cheyenne, WY 82002
(307) 777-7788
(307) 777-3508 (fax)

## NATIONAL COUNSELOR CERTIFICATION

National Board for Certified Counselors
3 Terrace Way, Ste. D
Greensboro, NC 27403
(910) 547-0607
(910) 547-0017 (fax)
nbcc@nbcc.org (email)
www.nbcc.org

# PROFESSIONAL ASSOCIATION

American Counseling Association
5999 Stevenson Ave.
Alexandria, VA 22304-3300
(703) 823-9800
1-800-347-6647
(703) 823-0252 (fax)
www.counseling.org

# HONORARY SOCIETY

Chi Sigma Iota Counseling Academic and Professional Honor Society
International School of Education
250 Ferguson Bldg.
University of North Carolina at Greensboro
Greensboro, NC 27412-5001
(910) 334-4035
www.csi-net.org

# ACCREDITATION AGENCY

Council for Accreditation of Counseling and Related Educational Programs
5999 Stevenson Ave.
Alexandria, VA 22304
(703) 823-9800 ext.301
(703) 823-0252 (fax)
(703) 370-1943 (TTD)

Source: American Counseling Association

# APPENDIX G

# Commonly Used Abbreviations

| @ | about |
|---|---|
| AA | Alcoholics Anonymous |
| a.c. | before meals |
| A.C.S.W. | Academy of Certified Social Workers |
| ad. Iib. | as much as desired |
| adm. | admitted |
| adol. | adolescent |
| AMA | against medical advice |
| amt. | amount |
| a.o. | anyone |
| appt. | appointment |
| as. | of each |
| ASA | against staff advice |
| ASAP | as soon as possible |
| ASR | at staff request |
| B/4 | before |
| b/c | because |
| b.f. | boyfriend |
| b.i.d. | twice a day |
| b.i.n. | twice a night |
| bl. | black |
| B.M. | bowel movement |
| B.P. or B/P | blood pressure |

| | |
|---|---|
| bro. | brother |
| -c or w/ | with |
| c | canceled |
| Ca | calcium |
| CA | cancer |
| C.A.C. | Career Assessment Inventory |
| cap. | capsule |
| C.A.PS. | Career Ability Placement Survey Test |
| C.B.C. | complete blood count |
| cc | cubic centimeter |
| CCDC | Certified Clinical Drug Counselor |
| CCMHC | Certified Clinical Mental Health Counselor |
| cl or ct. | client |
| CNS | central nervous system |
| c/o | complains or complained of |
| C.P.T. | Common Procedural Terminology |
| C/R | cancel/reschedule |
| Cx | cancel |
| D or dos. | dose |
| D&A | drug and alcohol |
| D.A.T. | Differential Aptitude Test |
| d.c. | discontinue |
| Detox | detoxification |
| dis. | discontinued |
| DNA | did not appear |
| dr. or z | dram |
| Dr. | doctor |
| D.S.M. | Diagnostic and Statistical Manual |
| D.T.'s | delirium tremens |
| Dx | diagnosis |
| E | evolved |
| ECT | electroconvulsive therapy |
| EEG | electroencephalogram |
| e.g. | for example |
| EKG or ECG | electrocardiogram |
| e.o. | everyone |
| ER, EW, or EMR | emergency room or emergency ward |
| exam. | examination |
| fa. | father |
| fl. or fld. | fluid |
| F.R. | fully resolved |
| ft. | feet |
| G.A. | Gamblers Anonymous |
| g.f. | girlfriend |
| G.I. | gastrointestinal |
| gm. | gram |

| | |
|---|---|
| gma. | grandmother |
| gmpa. | grandfather |
| gr. | grain |
| GRP | group |
| gtt. | drop |
| H & P | history and physical exam |
| h.s. | hour of sleep or bedtime |
| ht. | height |
| H2O | water |
| Hx | history |
| ICDM 10 | International Classification of Diseases Manual (Vol. 10) |
| i.e. | that is |
| Iden. | identification |
| I.M. | intramuscular |
| in. | individual |
| info. | information |
| I.S.B. | Incomplete Sentences Blank |
| I.V. | intravenous |
| J.C.A.H.O. | Joint Commission on Accreditation of Healthcare Organizations |
| K | potassium |
| L. | left |
| Lge. | large |
| liq. | liquids |
| L.P.C. | Licensed Professional Counselor |
| L.P.C.C. | Licensed Professional Clinical Counselor |
| L.P.N. | Licensed Practical Nurse |
| mcg. | microgram |
| m Eq. | milliequivalent |
| Mg | magnesium |
| mg. | milligram |
| Mg S04 | magnesium sulfate |
| min. | minute |
| ml. | milliliter |
| mm. | millimeter |
| M.M.P.I. | Minnesota Multiphasic Personality Inventory |
| mo. | mother |
| mod. | moderate |
| mot. | motivation |
| MS | mental status |
| Mtg. | meeting |
| N.A. | Narcotics Anonymous |
| NA | not applicable |
| Na. | sodium |
| n/ach | not achieved |

| | |
|---|---|
| N.C.C. | National Certified Counselor |
| neg. | negative |
| n.o. | no one |
| no. or # | number |
| ns | no show |
| NPO | nothing by mouth |
| N & V | nausea and vomiting |
| obs. | observation |
| O.B.S. | organic brain syndrome |
| o.d. | everyday |
| O.D. | overdose |
| oint. | ointment |
| o.m. | every morning |
| OTO | one time only |
| oz. or z | ounce |
| p | present |
| p.c. | after meals |
| pil. | pill |
| PO. | probation officer |
| p.o. | by mouth |
| P.R. | partially resolved |
| prn | whenever necessary, as needed |
| pt. | patient |
| q. | every |
| q.d. | every day |
| q.h. | every hour |
| q.i.d. | four times a day |
| q.o. | every other |
| q.o.d. | every other day |
| q.s. | quantity sufficient |
| q.2h. | every two hours |
| q.4h. | every four hours |
| R. or R | right |
| re: | regarding |
| rec. | received |
| R.N. | Registered Nurse |
| R/O | rule out |
| Rx | prescription, medication |
| s | without |
| s.c. | subcutaneously |
| schiz. | schizophrenia |
| sib. | sibling |
| sig. | label it |
| sis. | sister |
| S.O. | significant other |
| s.o. | someone |

| | |
|---|---|
| sol. | solution |
| s.o.s. | if necessary |
| ss | one half |
| stat. | immediately |
| sx | symptom |
| tab. | tablet |
| tbsp. | tablespoon |
| T/C | telephone call |
| temp. | temperature |
| t.i.d. | three times a day |
| t.i.n. | three times a night |
| T.O. | telephone order |
| T.P.R. | temperature, pulse, respiration |
| tr. | tincture |
| tsp. | teaspoon |
| Tx | treatment |
| U. | unit |
| U.R. | utilization review |
| UR | unresolved |
| V.O. | verbal order |
| w (or c) | with |
| w/i | within |
| wky. | weekly |
| w/o | without |
| wt. | weight |
| x | times |
| ∴ | therefore |
| ↑ | increase |
| ↓ | decrease |
| ♂ | male |
| ♀ | female |

# APPENDIX H

# Sample Intake Form

Client Name_____ Counselor Name_____

Date_____ Length of Interview_____

I. Client and Concern Description: *Often the formula denoting age, race, gender, marital status in adults, followed by the words "complaining of . . ." or "reporting . . ."; for example, a 54-year-old Caucasian male, married, complaining of . . . or reporting . . .*

II. Psychosocial History: *Note here relevant historical data on the client, exploring such areas as childhood and adolescence, current and family of origin history (including children), legal history, medical history, religious history, employment history, educational history, military history, history of mental health contacts, etc.*

III. Mental Status: Client appeared alert (y,n) and oriented (y,n)

Cognitions: *The counselor explores current functioning in the cognitive arena, incl. memory, intellectual level, concrete v. abstract thinking abilities, evidence of hallucinations, delusions, etc.*

Affect: *Here the counselor notes current type, level, and intensity of emotions, incl. depression, anxiety, mania, etc.*

Behaviors: *In this area of mental status, the counselor takes special notice of any behavioral anomalies, incl. tics, psychomotor agitation or retardation, pressured speech, unusual gestures, etc.*

Risk of Harm to Self or Others: *This area of current functioning is of special import and, thus, is listed separately. The counselor needs to determine level of lethality and any and all ideation, plans, and means. We strongly suggest that the intern become familiar with agency and ethical policies in this area.*

IV.  Diagnostic Impression (DSM IV)

| Axis I | Clinical Syndromes |
| | Other Conditions That May Be a Focus of Clinical Attention |
| Axis II | Personality Disorders |
| | Mental Retardation |
| Axis III | General Medical Conditions |
| Axis IV | Psychosocial and Environmental Problems |
| Axis V | Global Assessment of Functioning |

V.  Treatment Recommendations: *As with the diagnostic impression, treatment recommendations are conclusional data resulting from the process of the interview. And as with the diagnostic impression, these recommendations are tentative and subject to modification as needed. They may include recommendations for psychological testing; psychiatric referral for medication assessment; referral to another professional; a specific type of therapy, such as marital, family, individual, group; a specific modality, such as hypnosis, biofeedback, cognitive/behavioral; hospitalization; or even no treatment. A treatment plan (see Appendix I) usually follows. We suggest that the client be as involved in the process as possible.*

VI.  Additional Remarks:

# APPENDIX I

# Sample Treatment Plan

Client Name _____

Counselor Name _____

Date _____

| Problem/ Concern | Goal | Treatment Methods | Expected Date of Achievement | Results | Follow-up |
|---|---|---|---|---|---|
| **1.** depressive symptoms | learn to i.d. auto negative thoughts | cognitive-behavioral strategies | 01/2001 | | |
| **2.** list others here | | | | | |
| **3.** | | | | | |
| **4.** | | | | | |
| **5.** | | | | | |

(This treatment plan follows a management by objective format. Note that it is specific, time-limited, and concise. Further, it allows for client input in the therapy process. Notice, too, that the goal is usually the inverse of the problem statement. We recommend that the client signs the form along with the counselor; thus it becomes a social contract. In the case of minors, we have them and their parents sign, allowing for some concrete investment in therapy by both minor and parent. Finally, we have discovered that managed care companies like such a tangible document.)

_____ _____ _____ _____
Client           Intern           Supervisor       Parent (as necessary)

# APPENDIX J

# Therapy Session Case
# Note Components

1. Type of treatment (e.g., Ind. TX)

2. Length of session (e.g., 1 hr.)

3. Mini Mental Status Examination (standard formula, e.g., "Ct. appeared alert, oriented")
   Hint: Note any aberrations in cognition, affect, and behavior.

4. What was reported? This is information reported by the client.
   Hint: Use ct.'s quotes as appropriate

5. What was discussed? This is information obtained during the session and is reported like "minutes of the meeting." Again, ct.'s quotes may be helpful.

6. Give homework if possible.
   Hint: Consider small, discrete tasks to help ensure successes
   Hint: To be completed by next session

7. Date/time of next session:

8. Clinician's signature; degree; license; credential (e.g., Sally Smith, MA, LPC, NCC)

Nota bene:

- One clinician we know of reviews notes with the ct. as appropriate in an effort to establish a collaborative (horizontal v. vertical) relationship. However, some agency policies do not allow for this.
- If client is suicidal, but not imminently so, you may want to consider inserting a statement (in essence, a behavioral contract) to this effect:

    "I will not try to kill myself without first calling the emergency # and/or clinician."

| | | | |
| --- | --- | --- | --- |
| Client signs | Counselor signs | Supervisor signs | Date |

Give a copy of this section to the client.

Insert direct quotes from the client to support this clinical stance.

Always follow policies and procedures of the agency, provided they are legal and ethical.

# APPENDIX K

# Sample Evaluation
# of Intern Form

Intern's name _____

Date of this evaluation_____ Name of agency _____

Agency address _____

Agency phone/ e-mail address _____

Univ. supervisor_____ Agency supervisor _____

Title _____ Title_____

Professional degree _____ Professional degree _____

Licensed as_____ No. _____ Licensed as_____ No. _____

Certified as_____ No. _____ Certified as_____ No. _____

Internship dates __/__/__ to __/__/__ Intern paid? _____

Total number of placement hours _____

Total number of supervisory hours _____

## Suggested Competencies for Interns

4 = Outstanding   3 = Good   2 = Fair   1 = Poor   NA = Not applicable

I. Communication skills
   a. Verbal skills _____
   b. Writing skills _____
   c. Knowledge of nomenclature _____
   Comments:

II. Interviewing
   a.  Structure of interview _____
   b.  Attending behaviors _____
   c.  Active listening skills _____
   d.  Professional attitude _____
   e.  Interviewing techniques _____
   f.  Mental status evaluation _____
   g.  Psychosocial history _____
   h.  Observation _____
   i.  Use of questions _____
   j.  Reflection _____
   k.  Empathy _____
   l.  Respect for differences _____
   Comments:


III. Diagnosis
   a.  Knowledge of assessment instruments _____
   b.  Knowledge of current DSM _____
   c.  Use of records _____
   d.  Ability to formulate a preliminary diagnosis _____
   Comments:


IV. Treatment
   a.  Ability to draw up a treatment plan _____
   b.  Ability to perform individual counseling _____
   c.  Ability to perform marital counseling _____
   d.  Ability to perform conjoint counseling _____
   e.  Ability to perform family counseling _____
   f.  Ability to perform group counseling _____
   g.  Crisis intervention skills _____
   h.  Ability to deal with various populations _____
   i.  Ability to make progress notes _____
   Comments:

V. Case management
   a. Knowledge of agency programs and professional staff roles _____
   b. Knowledge of community resources _____
   c. Discharge planning _____
   d. Follow-up _____
   e. Record keeping of client management _____
   Comments:


VI. Agency operations and administration
   a. Knowledge of agency mission and structure _____
   b. Awareness of roles of administrative staff _____
   c. Knowledge of agency goals _____
   d. Understanding of agency care standards _____
   Comments:


VII. Professional orientation
   a. Knowledge of counselor ethical codes _____
   b. Knowledge of agency professional policies _____
   c. Ability of intern to seek and accept supervision _____
   Comments:


Please write a brief summary statement of the intern as a future counselor.


Intern _____

Agency Supervisor _____

University Supervisor _____

Date _____

# APPENDIX L

# Sample Intern Evaluation of Site/Supervisor Form

Intern's Name_____

Date of This Evaluation _____ Name of Agency _____

Agency Supervisor _____ Title & License _____

Internship Dates __/__/__ to __/__/__ Total # Hours _____

## THE SITE

4 = Outstanding   3 = Good   2 = Fair   1 = Poor   NA = Not Applicable

   I. Overall agency operations and administration _____

  II. Introduction to agency mission and structure _____

 III. Awareness of roles of administrative staff _____

 IV. Knowledge of agency goals _____

  V. Understanding of agency care standards _____

 VI. Policies on duty to risk management (incl. duty to warn) _____

VII. Policies on confidentiality of records _____

Comments:

## THE SUPERVISOR

4 = Outstanding   3 = Good   2 = Fair   1 = Poor   NA = Not Applicable

Noting the competency areas suggested for my internship (refer to Evaluation of Intern Form and Chapter Two), I might rate my supervisor's assessment and/or training of me in these areas as:

    I.  Communication skills _____

   II.  Interviewing _____

  III.  Diagnosis (incl. knowledge of DSM-IV) _____

  IV.  Treatment (incl. treatment planning & termination) _____

   V.  Case management _____

  VI.  Agency operations and administration _____

 VII.  Professional orientation _____

VIII.  Knowledge and application of professional ethics _____

  IX.  Ability to process my professional issues with supervisor _____

Comments:

# Index

TO THE OWNER OF THIS BOOK:

We hope that you have found *The Counselor Intern's Handbook,* Second Edition, useful. So that this book can be improved in a future edition, would you take the time to complete this sheet and return it? Thank you.

School and address: ———————————————————————————

Department: ————————————————————————————————

Instructor's name: —————————————————————————————

1. What I like most about this book is: ——————————————————

———————————————————————————————————————

———————————————————————————————————————

2. What I like least about this book is: —————————————————

———————————————————————————————————————

———————————————————————————————————————

3. My general reaction to this book is: ——————————————————

———————————————————————————————————————

4. The name of the course in which I used this book is: ————————

———————————————————————————————————————

5. Were all of the chapters of the book assigned for you to read? ————

    If not, which ones weren't? ——————————————————————

6. In the space below, or on a separate sheet of paper, please write specific suggestions for improving this book and anything else you'd care to share about your experience in using the book.

———————————————————————————————————————

———————————————————————————————————————

———————————————————————————————————————

———————————————————————————————————————

———————————————————————————————————————

Optional:

Your name: _____ Date: _____

May Brooks/Cole quote you, either in promotion for *The Counselor Intern's Handbook,* Second Edition, or in future publishing ventures?

Yes: _____ No: _____

Sincerely,

*Christopher Faiver*
*Sheri Eisengart*
*Ronald Colonna*

---

FOLD HERE

- - - - - - - - - - - - - - - - - - - - - - - - - - - - - - - - - - - - - -

NO POSTAGE
NECESSARY
IF MAILED
IN THE
UNITED STATES

## BUSINESS REPLY MAIL
FIRST CLASS        PERMIT NO. 358        PACIFIC GROVE, CA

POSTAGE WILL BE PAID BY ADDRESSEE

ATT: *Eileen Murphy, Counseling Editor* _____

**BROOKS/COLE PUBLISHING COMPANY**
**511 FOREST LODGE ROAD**
**PACIFIC GROVE, CALIFORNIA 93950-5098**

FOLD HERE